Results from the
Second Mathematics Assessment
of the
National Assessment of
Educational Progress

THOMAS P. CARPENTER
MARY KAY CORBITT
HENRY S. KEPNER, JR.
MARY MONTGOMERY LINDQUIST
ROBERT E. REYS

MARY KAY CORBITT, *Editor*

NATIONAL COUNCIL
OF TEACHERS OF MATHEMATICS

Copyright © 1981 by
THE NATIONAL COUNCIL OF TEACHERS OF MATHEMATICS, INC.
1906 Association Drive, Reston, Virginia 22091

Library of Congress Cataloging in Publication Data:

Main entry under title:
Results from the second mathematics assessment of
the National Assessment of Educational Progress.
Bibliography: p.
1. National Assessment of Educational Progress
(Project). Mathematics objectives.   I. Carpenter,
Thomas P.   II. Corbitt, Mary Kay.   III. National
Council of Teachers of Mathematics.
QA13.R47   510′.7′1073   81–4322
ISBN 0–87353–172–8   AACR2

This project was supported by a grant (SED 7920086) from the National
Science Foundation to the National Council of Teachers of Mathematics. Any
opinions, findings, and conclusions or recommendations expressed in this publication are those of the authors and do not necessarily reflect the views of the
National Science Foundation.

Printed in the United States of America

Table of Contents

Preface

The National Assessment of Educational Progress completed its second mathematics assessment during the 1977-78 school year. This monograph, prepared by a team of members of the National Council of Teachers of Mathematics, represents a comprehensive discussion of the results of that assessment.

NCTM has worked cooperatively with National Assessment in the past. In 1974, an NCTM committee was appointed (James W. Wilson, chairman, Robert E. Reys, Terrence Coburn, and Thomas P. Carpenter) to prepare interpretive reports of the first mathematics assessment. Articles subsequently appeared in the Mathematics Teacher, the Arithmetic Teacher, and the Journal for Research in Mathematics Education, and a comprehensive monograph was published by the Council. These articles provided the only interpretive reports of NAEP results available to classroom teachers.

The interpretive reporting of the results of the first mathematics assessment was well received; this prompted NCTM to continue its interest in providing the service of producing and making such reports available. The NCTM Board of Directors, on the recommendations of three NCTM committees, expressed its support for the preparation of interpretive reports of the results of the second mathematics assessment of NAEP.

A committee composed of James W. Wilson (representing the Committee on the Mathematics Curriculum for the 1980s), Elizabeth Fennema (Research Advisory Committee), Henry S. Kepner, Jr. (Instructional Affairs Committee), and Jane Armstrong (National Assessment of Educational Progress) developed a proposal that was ultimately funded by the National Science Foundation to support an NCTM interpretive project. Project team members were Thomas P. Carpenter (project director), Mary Kay Corbitt, Henry Kepner, Mary Montgomery Lindquist, and Robert E. Reys. The team has prepared interpretive reports of NAEP mathematics data for several publications, including the Arithmetic Teacher and the Mathematics Teacher.

This monograph represents the most complete interpretive discussion of the data from the second mathematics assessment and also serves to highlight changes in mathematical achievement that have occurred since the time of the first mathematics assessment. Individual team members authored the separate chapters of the monograph; Constance Anick, Thomas Carpenter, and Carol Smith prepared Appendix A, a report of results by race, and Elizabeth Fennema and Thomas Carpenter prepared the report of the results by sex, Appendix B. Mary Kay Corbitt compiled and edited the volume, with the entire team assuming final editorial responsibility.

Thanks are expressed to National Assessment for their cooperation with the project and to Max Sobel, NCTM president, for preparing the introductory remarks to this monograph.

Introductory Remarks

The results of the second mathematics assessment of the National Assessment of Educational Progress (NAEP) provided the mathematics community with a measure of its standing at the beginning of the 1980s. After a decade of curriculum reform, followed by the back-to-basics movement of the 1970s, the NAEP results suggest strong implications for the mathematics curriculum of the 1980s that mathematics educators need to evaluate with care.

The decline in performance between the first assessment in 1973 and the second one in 1978 gave rise to headlines that appeared to be damaging to teachers of mathematics. The results clearly indicated that the emphasis on basics has produced a generation of students who can compute but cannot apply their knowledge to the solution of real-life problems. However, it is important to recognize that there is no need for the mathematics profession to be defensive about the NAEP results. Rather, we need to make use of these results in a positive and constructive manner as we enter the change process for curriculum reform in the 1980s.

The results of the second mathematics assessment relate directly to the NCTM recommendations for school mathematics of the 1980s as stated in An Agenda for Action. The very first recommendation in this document states that "problem solving must be the focus of school mathematics in the 1980s." The NAEP results strongly emphasize the importance of this major concern and lend support to this fundamental recommendation.

It is interesting to note that the NAEP interpretive panel concluded their study by offering a number of specific recommendations, one of which dealt with teacher education (NAEP Mathematical Applications 1979, p. 45):

> There is a need for more consistent and comprehensive teacher edution in the mathematics area. More attention should be devoted to preparing teachers to teach concepts and problem solving so they will not tend to rely as heavily on drill and memorization work.

It is apparent that an adequate number of certified and qualified teachers of mathematics is essential if we are to implement the recommendations of the NAEP panel and those found in An Agenda for Action. The current shortage of teachers of mathematics is of grave concern to all professional educators and must be made the concern of society as well. The NAEP results point up the need for curriculum reform and lend credence to the urgent need for a greater number of fully certified teachers of mathematics.

We are indebted to the NAEP/NCTM interpretive team for preparing this monograph on the results from the second mathematics assessment of the National Assessment of Educational Progress. It is particularly fortunate that the team has chosen to relate the results directly to the NCTM recommendations for the mathematics curriculum of the 1980s. Thus this document should serve the profession well as we work with teachers and with the general public to interpret the results of the mathematics assessment and to promote the important recommendations found in An Agenda for Action.

Max A. Sobel

1

Overview of National Assessment

When the U.S. Office of Education was chartered in 1867, one charge to its commissioners was to determine the nation's progress in education. The National Assessment of Educational Progress (NAEP) was initiated a century later to address that charge in a systematic way.

Each year since 1969, National Assessment has gathered information about levels of educational achievement across the country and reported its findings to the nation. NAEP surveys the educational attainments of 9-year-olds, 13-year-olds, and 17-year-olds in ten learning areas: art, career and occupational development, citizenship, literature, mathematics, music, reading, science, social studies, and writing. Different learning areas are assessed every year, and all areas are periodically reassessed in order to measure possible changes in educational achievement. National Assessment has interviewed and tested more than 810 000 young Americans since 1969.

Goals of the National Assessment

National Assessment provides information to education decision makers and practitioners that can be used to identify education problem areas, to establish education priorities, and to determine the national progress in education. To do so, National Assessment must remain flexible enough to accommodate possible extensions, refinements, and modifications. The following goals have been established for the project by the National Assessment Policy Committee, the Analysis Advisory Committee, and the National Assessment staff:

Goal 1: To measure change in the educational attainments of young Americans

Goal 2: To make available on a continuing basis comprehensive data on the educational attainments of young Americans

Goal 3: To use the capabilities of National Assessment to conduct special-interest probes into selected areas of educational attainment

Goal 4: To provide data, analyses, and reports understandable to, interpretable by, and responsive to the needs of a variety of audiences

Goal 5: To encourage and facilitate interpretive studies of NAEP data, thereby generating implications useful to education practitioners and decision makers

Goal 6: To facilitate the use of NAEP technology at state and local levels when appropriate

Goal 7: To continue to develop, test, and refine the technologies necessary for gathering and analyzing NAEP achievement data

Goal 8: To conduct an ongoing program of research and operational studies necessary for the resolution of problems and refinement of the NAEP model (Implicit in this goal is the conduct of research to support previous goals.)

Portions of chapters 1 and 2 were taken directly from NAEP sources. Thanks are expressed to National Assessment for permission to use this material.

Methodology

To measure the nation's educational progress, National Assessment estimates the percentage of respondents (at any of three age levels) who are able to answer a question or perform a task acceptably. Each question or task (called an exercise) reflects an education goal or objective. The exercises are administered to scientifically selected samples that take into account such variables as size and type of community, race, and geographic region. The General Information Yearbook of National Assessment provides a complete description of the methodology employed in the assessment.

Sample. The exercises in the NAEP mathematics assessment were administered to a carefully selected representative sample of 9-, 13-, and 17-year-olds. Since NAEP reports the percentages of students who responded correctly to an exercise, it is not necessary for each student to respond to every exercise. Rather, an item-sampling procedure was used so that each respondent received only about a tenth of the exercises administered at each age level. So although over 70 000 students participated in the assessment, each exercise was administered to approximately 2400 respondents at each age level. Altogether, approximately 230 exercises were administered to 9-year-olds, 350 to 13-year-olds, and 430 to 17-year-olds.

Since respondents are selected by age, students from several different grades were included at each age level. Approximately 25 percent of the 9-year-olds were in grade 3, and most of the remaining 75 percent were in grade 4. Approximately 27 percent of the 13-year-olds were in grade 7, and 69 percent were in grade 8. The 17-year-old sample consisted only of 17-year-olds who were still in school. Since approximately 12 percent of the 17-year-olds are not in school, the sample at this age level includes a somewhat different sample of the population than is represented in the other two age groups. Of the 17-year-olds included in the assessment, about 72 percent were in grade 11, 14 percent in grade 10, and 14 percent in grade 12.

Administration procedures. All exercises were administered by specially trained exercise administrators to groups of fewer than 25 students in 45-minute testing situations. To standardize procedures and minimize reading difficulty, all exercises were presented on a paced audiotape as well as in exercise booklets. The 9-year-olds were assessed during January and February 1978, the 13-year-olds during October and November 1977, and the 17-year-olds during March and April 1978.

Assessment exercises and results. The concept of a total score for an individual is inappropriate for National Assessment data. Rather, exercises are designed to be interpreted on an exercise-by-exercise basis. Both multiple-choice and open-ended exercises were included. Scoring guides were developed for the open-ended exercises so that the percent of respondents making specific errors could be identified.

Reports on the results of the second mathematics assessment are available from NAEP headquarters at Suite 700, 1860 Lincoln Street, Denver, CO 80295. Reports include Mathematical Knowledge and Skills, Mathematical Understanding, Mathematical Applications, and Changes in Mathematical Achievement: 1973-1978. Results for all released exercises (The Second Assessment of Mathematics, 1977-78: Released Exercise Set) and Mathematics Objectives: Second Assessment, which describes the development of mathematics objectives and exercises are also available from NAEP. Much of the material presented in chapters 1 and 2 of this volume was drawn from the objectives booklet.

A complete listing of available references and reports of the data from the second mathematics assessment follows the appendixes. Selected references to the first mathematics assessment results are also included.

2

The Second Assessment of Mathematics

Framework for Objectives and Exercises

The objectives that guided the development of exercises for the assessment were selected by panels of mathematicians, mathematics educators, classroom teachers, and interested lay citizens to reflect important goals of the mathematics curriculum. These groups concluded that the mathematics curriculum should be concerned with a broad range of objectives. Accordingly, the assessment focused on five major content areas: (1) numbers and numeration, (2) variables and relationships, (3) geometry (size, shape, and position), (4) measurement, and (5) other topics, which included probability and statistics, graphs and tables, computer literacy, and attitudes toward mathematics. Each content area was assessed at four levels: (1) knowledge, (2) skill, (3) understanding, and (4) application. The scheme for developing objectives and exercises can be represented as shown in Figure 2.1.

PROCESS \ CONTENT	A. Numbers and Numeration	B. Variables and Relationships	C. Shape, Size and Position	D. Measurement	E. Other Topics
I. Mathematical knowledge					
II. Mathematical skill					
III. Mathematical understanding					
IV. Mathematical application					

Figure 2.1. Scheme for developing objectives and exercises.

Content

The content domain for the second assessment of mathematics drew primarily from the current curriculum of elementary and secondary schools, although some projection of future mathematics emphases was acknowledged (for example, greater use of metric measures, earlier introduction of decimals, and calculator computation). Since the assessment is not designed to provide different questions to students with different mathematics backgrounds, the exercises focus on content that has been studied by the majority of students at a given age level. Therefore, topics that would require a formal course in geometry, advanced algebra, or precalculus were not included. These five content categories were used in this assessment:

1. Numbers and numeration
2. Variables and relationships
3. Size, shape, and position
4. Measurement
5. Other topics

These content categories helped to organize the domain but were not intended to be represented equally in the assessment. The selection of exercises was based on judgments about the appropriateness for each age group and the relative emphasis in the school curriculum.

Numbers and numeration. This category contained the largest number of exercises because of its importance in the curriculum. Exercises dealt with the ways numbers are used, processed, or written. Knowledge and understanding of numeration and number concepts were assessed for whole numbers, fractions, decimals, and integers and percents, with considerable emphasis placed on the four basic operations. Number properties and order relations were also assessed. Problem-solving exercises included routine number problems, nonroutine problems, and consumer problems. Nonroutine problems are exercises that are not normally encountered in the curriculum but are understandable to the age group. Consumer problems dealt primarily with the uses of mathematics in commercial situations (for example, buying and selling, interpreting graphs, and saving money) and were emphasized more at the 17-year-old level than at the two younger levels.

Variables and relationships. The use of variables and relationships corresponds to an important part of the school mathematics curriculum. The exercises for this content category dealt with facts, definitions, and symbols of algebra; the use of variables in equations and inequalities; the use of variables to represent elements of a number system; functions and formulas; coordinate systems; and exponential and trigonometric functions. Most of the exercises in this category were given only to 17-year-olds.

Size, shape, and position (geometry). The exercises in this content category measured objectives related to school geometry. The emphasis in the assessment was not on geometry as a formal deductive system. The exercises concerned plane and solid shapes, congruence, similarity, properties of triangles, properties of quadrilaterals, constructions, sections of solids, other basic theorems and relationships, and rotations and symmetry.

Measurement. A large portion of the assessment was devoted to measurement, reflecting an increased emphasis on measurement in the school curriculum. The exercises covered appropriate units; equivalence relations; instrument reading; length, weight, capacity, time, and temperature; perimeter, area, and volume; nonstandard units; and precision and interpolation. A substantial number of the measurement exercises required the use of metric units.

Other topics. Other mathematical content topics included in this assessment at all age levels were probability and statistics; graphs, tables, and charts; and logic. Special assessment exercises and procedures were developed to assess attitudes related to mathematics, computer literacy, and the use of the hand calculator.

Approximately 40 percent of the exercises dealt with number and numeration, with the remaining 60 percent distributed over the other four content areas.

Process

The process domain for the second assessment comprised four categories:
1. Mathematical knowledge
2. Mathematical skill
3. Mathematical understanding
4. Mathematical application

Like the content domain, the process domain could be used to classify either objectives of mathematics instruction or exercises to assess the learning of mathematics. Although each category suggests a type of mental process, neither objectives nor exercises fall neatly into a single process category--if only because the process has to be inferred and different people may use different processes or different combinations of processes. Arbitrary decisions must be made in using any system of process categories. Such a system is helpful, however, in ensuring a consideration of the diversity possible within a given content category.

Mathematical knowledge. Mathematical knowledge refers to the recall and recognition of mathematical ideas expressed in words, symbols, or figures.

Mathematical knowledge relies, for the most part, on memory processes. It does not ordinarily require other, more complex mental processes. Exercises that assess mathematical knowledge require that a person recall or recognize one or more items of information. An example of an exercise involving recall would be one that asks for a multiplication fact, such as the product of 5 and 2. Another example would be an exercise asking for the statement of a mathematical relationship, such as the law of cosines. An example of an exercise involving recognition would be one that presents several symbols and asks which symbol means <u>parallel</u>.

<u>Mathematical skill</u>. Mathematical skill refers to the routine manipulation of mathematical ideas. Exercises that assess mathematical skill assume that the required algorithm has been learned and practiced. They do not require a decision concerning which algorithm to use or that the algorithm be applied to a new situation. Such an exercise aims at measuring proficiency in carrying out the algorithm rather than the understanding of how or why it works. Mathematical skill is assessed by exercises that require the performance of such specified tasks as making measurements, multiplying two fractions, graphing a linear equation, or bisecting an angle.

<u>Mathematical understanding</u>. Mathematical understanding refers to the explanation and interpretation of mathematical knowledge. Mathematical understanding relies primarily on translation processes. Mathematical knowledge can be expressed in words, symbols, or figures, and the translation may be within or between any of these modes of expression. Mathematical understanding involves memory processes as well as processes of associating one item of knowledge with another.

Exercises that assess mathematical understanding require that an explanation or an illustration for one or more items of knowledge be provided. They require the transformation of knowledge but not the application of that knowledge to the solution of a problem. An example of an exercise involving explanation would be one that asks why a certain graph is not the graph of a function. Examples of exercises involving interpretation would be those that ask for a drawing of an array to represent 6 × 7 or that ask for an equation to represent the information in a word problem.

<u>Mathematical application</u>. Mathematical application refers to the use of mathematical knowledge, skill, and understanding. Mathematical application relies on processes of memory, algorithm, translation, and judgment.

Exercises that assess mathematical application require a sequence of processes that relate to the formulation, solution, and interpretation of problems. The processes may include recalling and recoding knowledge, selecting and carrying out algorithms, making and testing conjectures, and evaluating arguments and results. An exercise might require the solution of a standard problem on proportion, or it might require the demonstration that two geometric figures are congruent. It might require the estimation of the amount of carpet needed for a room or the formulation of a problem given a graph of statistical data. Exercises assessing mathematical application may vary from routine textbook problems to exercises dealing with mathematical arguments. In these exercises the person being tested is not told exactly what to do or how to do it. He or she must use reasoning and decision-making processes as well as mathematical knowledge, skill, and understanding.

Approximately 15 percent of the exercises were at the knowledge level, 25 percent at the skill level, 25 percent at the understanding level, and the remaining 35 percent at the application level.

Framework for Reporting Results

Background Variables

In addition to reporting national results, NAEP presents results on the basis of certain demographic variables. These include race, sex, parental education, size and type of community, and geographic region of the country. (See appendixes for discussion of race and sex data.) In addition to these variables, the second mathematics assessment gathered data on the mathematics courses that 17-year-olds had taken during their high school careers. The number of 17-year-olds who had completed at least half a year of selected mathematics courses is summarized in Table 2.1. These data indicate that there is substantial attrition in upper level mathematics courses. Very few students are graduating from high school with a sufficient background to begin studying calculus; this means that most high school students are graduating with an insufficient mathematics background to begin careers or majors in most scientific or social science fields.

Table 2.1

Mathematics Courses Taken by 17-Year-Olds

Course	Percent Having Completed at Least ½ Year
General or Business Mathematics	46
Prealgebra	46
Algebra 1	72
Geometry	51
Algebra 2	37
Trigonometry	13
Precalculus/Calculus	4
Computer Programming	5

Change

A major objective of National Assessment is to describe change in students' educational attainments. For this purpose certain exercises from each assessment are not released and are readministered in future assessments. Over a hundred exercises from the 1972-73 assessment were readministered during the 1977-78 assessment. The results of these exercises provide the basis for measuring change in students' performance in mathematics. Similarly, about two-thirds of the exercises from the second assessment are not released to provide measures of change in future years.

In order to provide valid comparisons, administration and scoring procedures were identical for both assessments, and corresponding age levels were tested during the same time of the year. A more complete discussion of change data is presented in chapter 3.

3

The Assessment of Change

One of the major objectives of National Assessment is to measure change in the educational achievements of 9-, 13-, and 17-year-old students. Data from exercises given during the first and second mathematics assessments give some indication of changes in students' mathematics achievement from 1973 to 1978. These data, along with a discussion of some of the problems associated with measuring changes in achievement, are the subject of this chapter.

In order to measure changes in performance, about two-thirds of the exercises for each assessment are not released so that they can be readministered in future assessments. All change data are based on this common set of exercises. National Assessment attempts to make testing conditions as alike as possible in each assessment. Corresponding age levels are tested during the same time of the school year. The amount of time students are given to respond to the individual change exercises is identical in each administration. Scoring procedures as well as administration procedures are also consistent between assessments.

In this assessment, the placement of the exercises within the testing booklets varies between assessments, but approximately 20 percent of the more than one hundred change exercises were multiple choice; the remainder were open ended. The 1973 responses to the open-ended change exercises were rescored using 1978 scoring guides to ensure comparable sets of responses.

Problems in Interpreting Change Results

Although every reasonable effort was made to make testing conditions similar for both administrations of the change exercises, some caution is necessary in interpreting the results. Even though testing conditions were held constant for individual change exercises, the change items were packaged with different sets of items in each assessment. Differences in difficulty or testing conditions of the nonchange exercises could have affected performance on the change exercises. Another complication is that the population of 17-year-olds may have undergone changes in the time between assessments. If more 17-year-olds are staying in school now than five years ago, this would create a different sample of 17-year-olds to be included in the assessment.

In addition to these potential threats to the validity of the change data, several other considerations must be kept in mind in interpreting the change results. Currently, change data exist for only two points in time; this does not provide a sufficient pattern of change to predict trends. The declines in performance observed from the first to the second assessment may not appear so obvious if a comparable increase in performance is observed from the second to the third assessment. By the same token, the declines would assume far greater importance if the third assessment results showed still further declines. The point to be kept in mind is that performance measured on two different groups at two different times can be reasonably expected to show small fluctuations. Only over a longer period of time do trends in levels of performance become apparent and more interpretable.

Another important consideration in determining what changes in perfor-

8

mance are significant involves the issue of statistical significance versus educational significance. NAEP reported that "change" had occurred if differences in levels of performance were (statistically) significant at the .05 level. In some instances, this meant that a change of 2 percentage points in performance levels was statistically significant. Mathematics teachers must question whether a change of this magnitude represents a change that is important from an educational viewpoint.

Finally, in interpreting the change results, it is important not to look for simplistic explanations for observed changes. There is a five-year interval between the two assessments, and students at a particular age level at one point in time may have been exposed to a wide range of experiences different from those of similar students five years earlier. To attempt to attribute any observed change to a single cause--for example, to changes in the mathematics curriculum--is overly simplistic.

Summary of Change Results

National Assessment exercises are designed to be interpreted on an exercise-by-exercise basis. In general, no attempt is made to create scales of exercises, and average scores for sets of exercises are difficult to interpret. To provide a simple summary of the change results, however, the average change over different aggregations of exercises was computed. Because these results have received such widespread attention in the popular press, we have summarized results for the basic process levels below. We want to point out, however, that a great deal is lost in such summary statistics and that these data are especially vulnerable to misinterpretation.

The set of exercises used to measure change was actually a small subset of the entire set of exercises administered. The set is not representative of the entire mathematics curriculum. Thus, there is an element of risk associated with attempting to generalize about students' performance across all areas of the curriculum and all process levels when the set of change exercises is considered.

Another difficulty with average scores is that averages are heavily influenced by extreme scores. For example, two exercises accounted for over three-fourths of the total decline in performance at age 9. These exercises involved the application of multiplication and division, which are first introduced in the third and fourth grades. These two exercises are hardly representative of the mathematics we would expect 9-year-olds to have learned, but they account for most of the change in performance.

Another limitation in the use of average scores is that they suggest a uniform decline in performance and hide shifts in student performance that may be meaningful. For example, on the topic of whole number computation, a blanket statement that "performance for 9-year-olds remained relatively unchanged" hides the fact that there were increases of around 5 percentage points on some addition and multiplication exercises, and decreases of that magnitude on other addition and subtraction exercises. Thus, aggregated change data are potentially misleading, and one should certainly be cautious in attributing particular meaning to average change scores.

Table 3.1 summarizes changes in average performance levels between assessments for all age groups. At age 17, performance declined significantly at all process levels except knowledge. At age 13, there was a significant decline on skill and applications exercises, although the declines were less than those shown by the 17-year-olds. At age 9, the only significant decline was at the application level.

Before these results are discussed further, it should be noted that the patterns of change results are similar to those in other subject areas. Seventeen-year-olds have shown declines in performance in several areas including writing, social studies, and science, whereas the 9-year-olds' levels of performance have generally remained unchanged in other subjects. Just as they did on mathematics, the 13-year-olds showed slight declines in some other subject areas, but those declines were less than those of the 17-year-olds.

Table 3.1

Changes in Average Performance from 1973 to 1978

Process Level	Age	Number of Exercises	Average Performance Level (1978)	Percent Change**
Knowledge	9	17	55	−1
	13	16	64	0
	17	18	63	0
Skill	9	21	26	0
	13	37	49	−2*
	17	46	50	−5*
Understanding	9	***	--	--
	13	12	50	−2
	17	13	58	−4*
Application	9	9	32	−6*
	13	12	38	−3*
	17	25	29	−4*
All Exercises	9	55	37	−1
	13	77	51	−2*
	17	102	48	−4*

*Significant at .05 level
**Negative number indicates performance lower in 1977-78 than in 1972-73
***Not enough exercises administered to make a mean percent interpretable

Knowledge. As a group, the exercises in this category showed the smallest changes in performance levels between assessments. At the same time, the largest changes in performance on any exercises in the assessment were also contained in this category. These exercises dealt with students' knowledge of metric measures. On one exercise, 63 percent of the 13-year-olds selected the kilometer as the longest unit of measure from among a group of metric measures; this was an improvement of 26 percentage points over the first assessment. Other changes in performance on metric exercises showed improvements of 12 to 15 percentage points for both 13- and 17-year-olds.

Skill. The exercises in this category assessed students' ability to use algorithms or to manipulate symbols. Change results on the whole number computation exercises showed that performance levels of 9- and 13-year-olds remained relatively unchanged overall (declines were offset by gains), while the 17-year-olds showed some decline in their ability to deal with more complex calculations. There were significant declines at all ages on those change exercises that dealt with computation with fractions and percents.

Understanding. This process level was not emphasized during the first assessment, as reflected by the small number of change exercises in the category. The exercises covered several areas of content and are discussed more completely in the appropriate chapters.

Application. The overall decline in performance was most pronounced at the application level. Relatively few change exercises were given to the 9- and 13-year-olds, and results should be considered accordingly. Nonetheless, the declines were consistent across all age levels, and the overall levels of performance on application exercises were low.

Conclusion

The overall decline in performance observed from the first to the second mathematics assessment has received a good deal of coverage in the popular press. Most of this coverage has emphasized the negative aspects of the decline. The purpose of this chapter has been to highlight some factors that should be considered as change data are discussed. Although overall performance declines are in evidence, these should be weighed against the points raised in this chapter.

Furthermore, to fully understand the change results, it is necessary to examine the assessment result on an exercise-by-exercise basis. The pattern of change within each of the content areas is discussed in the following chapters. It is often difficult, however, to understand the pattern of change even where exercises are examined on an individual basis. Some exercises showed inconsistent patterns of change across age groups. For example, on one exercise given to all three age groups, respondents were shown pictures of two measuring cups, each containing an indicated amount of liquid, and were asked how much liquid there was in both cups. Performance levels for the 9- and 17-year-olds declined 1 and 2 percentage points, respectively, whereas performance for the 13-year-olds declined 13 percentage points. There is no obvious explanation for this drop in performance, and examination of results on other measurement and fractions exercises gives no indication of what may have caused the decline.

4

Number

Since number is central to much of the mathematics curriculum, this category received substantial emphasis on the assessment. About half the total number of exercises for 9-year-olds dealt with number. Of these approximately 120 exercises, three-fourths involved whole numbers and the rest involved fractions or decimals. The proportion of number exercises to other exercises was not quite as great for 13-year-olds as for 9-year-olds. There were, however, approximately 170 number exercises given to 13-year-olds, of which half involved whole numbers and the remainder dealt with fractions, decimals, percents, or integers. The approximately 150 number exercises represented about one-third of the total assessment for 17-year-olds.

Although number is often associated with computation, the scope of the assessment was much broader. The process levels of knowledge, skills, understanding, and applications were assessed for each type of number. Exercises assessing knowledge of number required simple recall, and those assessing skills involved computation with various types of numbers. Understanding of number was examined through exercises that required the student to translate from one form to another (e.g., write the number sentence to describe a model) or to explain steps in a procedure as well as through clusters of exercises. Application exercises included routine word problems, nonroutine problems, and consumer problems. Nonroutine problems were problems that are not normally encountered in a mathematics curriculum, and consumer problems dealt primarily with uses of mathematics in consumer situations.

The major results for all three age groups are reported here. It should be recognized that, to some extent, the number concepts and skills for 9-year-olds (approximately 25 percent in grade 3 and 75 percent in grade 4) are still developing, whereas the number ideas for the older two age groups should have reached a functional level. In other words, these exercises represent indexes of achievement for 9-year-olds and levels of retention for the older age groups.

This section is organized into subsections on whole numbers, fractions, decimals, percents, and other number topics. No attempt has been made to report all exercises, but representative exercises are given to support the major findings.

Whole Number

Overview of Results

This second mathematics assessment contained almost a hundred exercises that dealt with whole number concepts and properties, whole number operations, and word problems that used whole numbers. Although each of these categories is discussed separately, the major results are the following:

1. Most 9-year-olds are familiar with basic whole number concepts. Performance was generally high on exercises that involved counting, writing numbers from spoken words, ordering sets of numbers, and identifying ordinals. Nine-year-olds were less successful on tasks that directly involved place-value notions.

2. Performance for all ages was high on recall of basic facts and on simple calculations. The one exception was division; this operation presented difficulty to both of the older groups (the younger age was not assessed).
3. Computational skills showed growth from age to age and grade to grade. Often skills are not mastered at the time when greatest emphasis is given in the curriculum but at a later date after practice and application.
4. Performance was high on simple one-step problems for which the corresponding computational skills had been developed.
5. Performance on nonroutine problems and on multistep problems was generally poor.
6. There was not much change in performance from the previous assessment on those whole number exercises that did not involve problem solving or application; however, on the application level exercises there appeared to be some decline.

Whole Number Concepts and Properties

This section focuses on exercises that dealt with number and numeration concepts, order, properties, and number theory. Emphasis for the 9-year-olds was on number, numeration, and order, while for the 17-year-olds emphasis was on properties and number theory. Thirteen-year-olds were assessed in each category.

Number and numeration. The understanding of numbers and our numeration system is essential to any work with whole numbers. From the results of the thirteen number and numeration exercises, it was evident that 9-year-olds were able to count, to write numbers from the spoken word, and to identify ordinals. A selection of these exercises and their results appears in Table 4.1.

Table 4.1

Results of Number Exercises

Exercise	Percent Correct Age 9
Counting Exercise	
A. (Given a picture of 16 scattered squares.)** Count the squares. How many squares are there?	93
B. (Given a picture of 14 boxes.)** Here are 14 boxes of shoes. Each box has 2 shoes in it. Count by twos to find how many shoes are in the boxes.	81
C. (Given a picture of 13 boxes.)** Here are 13 boxes of marbles. Each box has 10 marbles in it. Count by tens to find how many marbles are in all the boxes.	82
Writing Exercise	
Write the number five hundred twenty-two.	88
Identifying Ordinal Numbers Exercise*	
Here is a picture of some trucks in line.** Fill in the oval under the fourth truck in the line.	95

*Unreleased exercise
**Pictures not included here

While the performance of 9-year-olds on the exercises in Table 4.1 was high, this was not true when place-value and grouping notions were essential, as in the exercise in Table 4.2. By age 13, students' performance on this exercise had risen to above the 80 percent level. Although the level of performance on the exercise in Table 4.2 was typical of other such exercises, there was a great contrast between it and one unreleased exercise in which 9-year-olds were asked to choose what the 6 stood for in a number such as 4617. On such an exercise, over 80 percent of the 9-year-olds responded correctly. The two tasks are almost logically opposite ones: given the place, state the digit, and given the digit, state the place. It could be argued that the word digit or the hundreds place caused the discrepancy in performance, but these results should also serve as a reminder that students may be able to do a task in one direction and not in the other.

Table 4.2

Results of a Place-Value Exercise

Exercise	Percent Responding Age 9	Age 13
A.　2079 What digit is in the tens place in the number in this box?		
○　2	9	3
○　0	16	8
●　7	56	82
○　9	9	5
○　I don't know	9	1
B.　23,486 What digit is in the thousands place in the number in this box?		
○　2	16	4
●　3	48	84
○　4	7	6
○　8	11	4
○　6	3	1
○　I don't know	12	1

One exercise asked 9-year-olds to estimate the number of birds in a picture. Approximately three-fourths chose the response 200, the closest estimate from the given choices. However, 21 percent chose 2000, which indicates a lack of feel for comparative sizes of quantities in the hundreds and thousands.

An unreleased exercise required the 9-year-olds to round a four-digit number to the nearest thousand. This is a skill that is only beginning to be developed at this age, and the percentage of correct responses (10 percent) indicated this level of development. This exercise and the number-writing exercise (Table 4.1) were two exercises also given on the previous assessment. There was little change in performance on either of them.

Order. The exercises discussed here deal with determining the relative

14

sizes of numbers (e.g., Which is larger, 24 or 37?), and describing this relationship with a mathematical sentence (24 < 37). They were administered only
to the two younger groups.

There were three unreleased exercises in which the 9-year-olds had to
choose the larger of two two-digit or three-digit numbers, put four two-digit
numbers in order, or choose the largest of four four-digit numbers. The most
difficult task was the latter, but three-fourths of the 9-year-olds were able
to choose the largest of four four-digit numbers. Choosing the larger of two
numbers was correctly done by over 95 percent of the 9-year-olds. The only
exercise also given to 13-year-olds was the one requiring four two-digit numbers to be put in order. Eighty-six percent of the 9-year-olds and 96 percent of the 13-year-olds were successful on this exercise. All in all, the
ordering of whole numbers presented little difficulty.

Several exercises used an inequality sign, such as <. When asked what
the symbol < meant, only 43 percent of the 9-year-olds and 67 percent of the
13-year-olds responded correctly. These results are shown in Table 4.3 along
with the percentages of incorrect responses. The most common distractor was
the phrase greater than. When the symbol was used in a mathematical sentence,
however, 73 percent of the 9-year-olds chose the true sentence. Although this
may be an indication that a symbol out of context does not always carry meaning, these results also make a strong case for the necessity of having students read mathematical symbols and sentences orally.

Table 4.3

Results of Exercises Involving the Inequality Sign

Exercise	Percent Responding	
	Age 9	Age 13
Symbol Only Exercise		
What does the symbol < mean? Fill in one oval.		
⟠ equal	2	1
● less than	43	67
⟠ greater than	45	31
⟠ I don't know	9	1
Symbol and Number Exercise		
Which of the following is TRUE?		
● 3 > 2	73	
⟠ 3 = 2	3	
⟠ 3 < 2	13	
⟠ I don't know	10	

Other exercises involving an inequality symbol required students to solve
open sentences, such as those of the multiple-choice exercises reported in Table 4.4. As can be noted from the results, most 9-year-olds chose the answer
to the related equality $5 + \square = 12$, and some 13-year-olds had the same tendency. Since the computation was simple, these results probably show a lack
of experience in handling open-order sentences. Or perhaps they indicate a
lack of acceptance of problems with no solutions or with more than one number
as the solution.

Table 4.4

Results of Open Inequality Exercises

Exercise	Percent Responding	
	Age 9	Age 13

First Inequality Exercise

$5 + \square < 12$

Which number(s) below can go in the $\square$
to make this number sentence true?

○ 7	54	20
● Any number less than 7	26	66
○ Any number greater than 7	12	11
○ I don't know	5	2

Second Inequality Exercise

For the following questions, select
the values for $\square$ from this set:

$\{1,3,5,7\}$

A. If $3 + \square > 5 + 1$, then $\square$ can be
equal to which of the following?

○ 3 only	13
○ 5 only	8
● 5 or 7	48
○ 7 only	14
○ None of them	10
○ I don't know	6

B. If $3 + \square > 5 + 5$, then $\square$ can be
equal to which of the following?

○ 3 only	2
○ 5 only	5
○ 5 or 7	7
○ 7 only	25
● None of them	54
○ I don't know	6

Properties. Ten exercises were administered that gave some indication of
9-, 13-, and 17-year-olds' ability to use certain properties of whole numbers.
Only two of the exercises were released; these appear in Table 4.5.

16

Table 4.5

Results of Exercises Involving Whole Number Properties

Exercise	Percent Correct		
	Age 9	Age 13	Age 17
Properties with Numbers Exercise			
As the following problems are read, decide what number goes in each box. Write only the ANSWERS in the space provided.			
A. $3 \times \square = 3$	79	91	96
B. $18 \times 0 \times 29 = \square$	35	49	67
C. $(37 \times 5) \times 2 = \square$	5	35	50
D. $0 \div 1000 = \square$	50	66	76
E. $5983 \times 83 = \square \times 5983$	20	63	76
F. $(18 + 12) + 8 = \square + 20$	6	45	65
Property with Letters Exercise			
a and b stand for whole numbers. If $a \times b = 84$, then $b \times a =$	70	--	--

Throughout the assessment, performance increased as age level increased. In general, no pattern of difficulty could be ascertained according to the type of property (e.g., commutative, associative). However, the additive properties seemed slightly easier than the multiplicative properties.

The relatively high percentage correct on part D should not be interpreted to mean that 9-year-olds have much understanding of this task. If they followed the pattern of responding to the question with one of the numbers in the exercise, as they so often did, then there was a good chance to respond "0" to the $0 \div 1000$. Similarly, the rather high percentage of correct responses to the last exercise in Table 4.5 can be slightly misleading if any conclusion is drawn about students' ability to express properties with letters. For the most part, exercises involving letters were more difficult than those involving numbers. Only 36 percent of the 13-year-olds and 52 percent of the 17-year-olds could choose an expression of a generalization of the associative property of addition.

Number theory. The majority of the exercises that could be classified as number theory dealt with even and odd numbers. A few asked about primes, common factors, or common multiples.

The only number-theoretic exercise given to 9-year-olds required them to give examples of odd and even numbers. Approximately 70 percent were successful on both tasks, with the odd numbers being slightly more difficult to generate than the even numbers.

The remaining nine exercises were given to the older groups, and generally about two-thirds were able to respond correctly. Table 4.6 contains representative exercises and results, although there were a few exceptions to this level of correct response.

One unreleased exercise asked 13- and 17-year-olds to find the least common multiple of three numbers. Only about one-fourth of the respondents gave the correct answer, with 30 percent of the 13-year-olds and 46 percent of the 17-year-olds giving a common factor as their answer.

Table 4.6

Representative Number Theory Exercises and Results

Exercise	Percent Responding	
	Age 13	Age 17

Even-Odd Number Exercise

If x and y are odd numbers, what is true about x + y?

○ It is odd.		8
● It is even.		58
○ It may be either even or odd depending on what x and y are.		31
○ I don't know		2

Prime Number Exercise

What is always true about any prime number?

○ It is less than 1000.	2	1
○ It is divisible by another prime number.	13	10
● It is divisible by only itself and 1.	58	71
○ It is a factor of 1.	16	12
○ I don't know	11	6

Operations with Whole Numbers

Thirty-four exercises dealt with the four operations of addition, subtraction, multiplication, and division of whole numbers. These included exercises concerned with concepts, basic facts, and computation. Word problems and open sentences that involved the four operations are discussed elsewhere.

Nine-year-olds were administered 31 of the 34 exercises and 13-year-olds were administered 24 exercises. These high proportions reflect the emphasis in the mathematics curriculum on whole number operations at ages 9 and 13, and the 15 exercises given to the 17-year-olds allow for comparisons across the three age levels.

Concepts. An introduction to the concepts underlying various operations is often made through the use of physical or pictorial models. Several exercises in this assessment dealt with pictorial models. The eight exercises given only to 9- and 13-year-olds were not released, and none of them were administered on the first assessment; therefore, no change data are available. A brief description of each exercise, along with the results, is given in Table 4.7. In each of the exercises, the students were asked to write or choose a sentence that represented a model, or to choose or complete a model to represent a sentence. In general, those models involving sets or arrays were easier than those involving the number line.

The results in Table 4.7 may reflect that the number line model is not used in all elementary mathematics programs. The incorrect responses to the four exercises involving the number line, however, showed a consistent pattern of misunderstanding. Students chose or wrote sentences or chose number lines that highlighted the numbers involved more often than they responded correctly. For example, if the model were the one following,

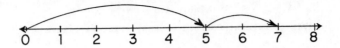

about twice as many 9-year-olds would respond "5 + 7" as would respond "5 + 2 = 7." Thirty-nine percent of the 13-year-olds made the same error. Similar errors were made with the subtraction, multiplication, and division models.

Table 4.7

Results of Exercises Involving Operation Concepts

Exercise	Percent Correct Age 9	Percent Correct Age 13
*Addition-set model exercise Choose sentence to represent set model of addition.	74	90
*Addition-number line exercise A. Write sentence to represent number line model of addition.	25	48
B. Complete number line model of addition to represent a sentence.	32	49
*Subtraction-number line exercise Write sentence to represent number line model of subtraction.	14	33
*Multiplication-additive meaning exercise Choose number line model to represent multiplication sentence.	33	48
*Multiplication-array exercise Write sentence to represent array model of multiplication.	67	
*Division-set model exercise Write sentence to represent set model of division.	26	52
*Division-number line exercise Write sentence to represent number line model of division.	22	24

*Unreleased exercise

If the number line is to be used in instruction, it must be used cautiously. Since the model does not seem to clearly suggest the operation, the meaning must be developed or the misunderstandings indicated here may occur. The mathematics curriculum should be constructed to ensure that students have a meaningful development of the basic operations. Certainly, many types of models can help this development, but they must be carefully selected and meaningfully taught.

Facts. The ability to recall basic facts, such as 3 + 4, 15 − 9, 6 × 8, and 15 ÷ 3, was evaluated on this second assessment but not on the first. Although there are no change data, the results from this assessment are positive. High performance levels can be seen by examining the average percent correct on the five or six facts from one exercise for each of the operations (see Table 4.8). The administration of the basic facts was done orally on a paced tape; no written fact was presented. Each exercise consisted of five or six facts, all of which were the same operation.

Table 4.8

Average Percent Correct on Basic Fact Exercises

Type of facts	No. of Parts	Average Percent Correct		
		Age 9	Age 13	Age 17
Addition	6	89	95	97
Subtraction	6	79	93	95
Multiplication	6	60	93	93
Division	5	*	81	89

*Division facts were not administered to 9-year-olds.

Performance improved from age 9 to age 13 on the addition facts and especially on the subtraction facts, even though drill decreases. The use of facts in other contexts probably helps to maintain and improve performance. Likewise, the slight increase from age 13 to age 17 is indicative of improved performance.

The average score for the 9-year-olds on multiplication facts was 60 percent correct. The 9-year-olds were tested in January and February and approximately one-fourth of them were in the third grade. On a similar exercise, data were available for 9-year-olds in the fourth grade, and their average was about 12 percentage points higher. Thus, there was improvement in basic fact multiplication skills from third to fourth grade, and this improvement continues with age.

Although the results on division facts were lower than the results on the multiplication facts, this may be reasonable, since multiplication facts are used more often. Even when dividing, say, 5924 by 8, it is common to ask, "Eight times what is something near 59?" When using calculators, the most likely way to check the reasonableness of an answer to a division problem is by multiplying. For both these reasons, it is probably a better strategy to emphasize the multiplication facts rather than the division facts.

Analyses were also made of what percentages of students responded correctly to all or some of the parts of any basic fact exercise. This analysis for the fact exercises given to all three age levels is reported in Table 4.9. If mastery of the facts is considered to be 80 percent (roughly five out of six or four out of five items), then these analyses show that 86, 73, and 43 percent of the 9-year-olds have mastered the addition, subtraction, and multiplication facts, respectively. The 13-year-olds mastered these same facts at levels of 94, 92, and 89 percent. The percentages of the 17-year-olds are slightly higher.

Table 4.9

Percentages of Correct Responses to Parts of Fact Exercises

	Addition			Subtraction			Multiplication			Division		
Ages	9	13	17	9	13	17	9	13	17	9	13	17
All six correct	71	86	92	55	77	82	29	78	76	*	**	**
5 or more correct	86	94	97	73	92	95	43	89	89	*	60	71
4 or more correct	91	96	98	81	95	97	54	94	96	*	78	78
3 or more correct	94	97	98	86	97	99	67	97	99	*	85	93
2 or more correct	95	98	99	90	98	99	79	98	99	*	90	96

*Division facts were not administered to 9-year-olds.
**Only five division facts were assessed in an exercise.

Two observations can be made from these data. First, there is a great change in the percentage from those who answered all six parts correctly to those who correctly answered five parts, but there is little change thereafter. Second, the percent correct for the individual parts of any one exercise did not vary greatly; that is, there was little variation in the difficulty of any fact. From these two observations it is reasonable to ask how often 100 percent can be reached on a timed test, even if mastery of all the facts is expected. While this may be a laudable goal, it may be one that different students need to reach at different times. Some students should not be held to mastery at the 100 percent level before being moved on to other topics.

At a time when many negative comments are made and many negative results are seen regarding students' achievement in mathematics, it is good to find a positive result. These results should be regarded in this light, but they should not lead to complacency. There still are questions that need to be answered about how to reach this level effectively or the desired level of mastery of basic facts.

Computation. All elementary mathematics programs spend a great amount of time on whole number computation, although 9-year-olds have worked little, if any, with multiplication and division computation. The older students should have mastered algorithms for all the operations. The results of computation exercises on this assessment show that the younger children have mastered simple addition, subtraction, multiplication, and division exercises. Because computational skills are of such interest, all the computation exercises used on the assessment will be presented or, if unreleased, will be described.

The results of the addition exercises are presented in Table 4.10. For the older two groups, the exercises presented little difficulty, so the discussion here will center on the 9-year-olds. Over 75 percent of the 9-year-olds could perform the simpler addition computations even when regrouping was required. For more complex addition involving four addends or four-digit addends, performance was at about the 50 percent level. Column addition provides greater chance for error than does the addition of two digits, but students also have less opportunity to practice this skill. When given a calculator, few 9-year-olds failed to respond to an addition exercise unless it was a column addition. Over one-fourth of them did not respond or responded "I don't know" to the column addition exercise (C) in Table 4.10 when a calculator was available. Although column addition was a problem for 9-year-olds, the older students were able to handle this task.

Table 4.10

Results of Addition with Whole Numbers

| Exercise | | Percent Correct | | |
		Age 9	Age 13	Age 17
Find the sums:	A.	90	98	98
A. 21 B. 37 C. 4285	B.	76	95	97
+54 +18 3273	C.	51	85	90
+5125				
What is the sum of 21 and 54?		76	83	--
Find the sum of 3 one- and two-digit numbers given horizontally.*		67	85	--

Add 43
      71
      75
    +92
		50	84	92

*Unreleased exercise

Finding the sum of 21 and 54 was presented in three ways. Two of the ways appear in Table 4.10 and the third was through a simple story problem. While 90 percent of the 9-year-olds could add the two numbers presented vertically, only 76 percent responded correctly to the exercise, "What is the sum of 21 and 54?" These results underscore that careful development of vocabulary is necessary. It is encouraging that the story problem carried the message of addition, as evidenced by the 82 percent of the 9-year-olds who responded correctly. Thirteen-year-olds had the same difficulty with the word sum as the 9-year-olds. When the word sum was used, their performance was about 15 percentage points lower than it was on either of the other two formats.

Subtraction has always been a more difficult operation for students than addition. The results of the second assessment confirm this, as can be seen by contrasting the results of the subtraction exercises in Table 4.11 with the addition results. This is especially true for the 9-year-olds. About 65 percent could perform simple subtraction computations involving regrouping, in contrast to 75 percent who could successfully add with regroupings. By the age of 13, over 90 percent of the students could subtract when regrouping was required.

Table 4.11

Results of Subtraction with Whole Numbers

Exercise	Percent Correct		
	Age 9	Age 13	Age 17
*Given in vertical notation, subtract with regrouping			
A. 2-digit numbers	66	92	94
B. 3-digit numbers, with 0's	44	81	88
C. 3-digit numbers	50	85	92
*Given in horizontal notation, subtract			
A. 3-digit numbers, no regrouping	74	96	97
B. 2-digit numbers, no regrouping	74	95	98
C. 1-digit number from 2-digit number, regrouping	58	92	96

*Unreleased exercise

All groups, however, had more difficulty subtracting when a zero was involved. The responses to such an exercise are given in Table 4.12. Responses to this exercise also typify the errors made by students when subtracting. For 9-year-olds, the most common error was a reversal error (subtracting the smaller digit from the larger digit in a column) and accounted for 17 to 32 percent of their errors. This was not true by age 13, and by age 17 this type of error was rarely found.

An unreleased multiple-choice exercise asked the students to estimate the difference between two four-digit numbers. Only 17, 52, and 69 percent of the 9-, 13-, and 17-year-olds chose the correct response. The most common incorrect response (from 39 percent at age 9 to 25 percent at age 17) was made by finding the difference between the two numbers in the thousands place and responding with that number of thousands. For example, if the students were asked to estimate the answer to 5116 - 2947, their most common incorrect response would be 3000.

Another exercise slightly removed from straightforward computation asked, "149 + 327 is how much greater than 149 + 320?" About one-half of the 9-year-olds and three-fourths of the 13-year-olds responded correctly. Although the computation involved is simple if approached as only looking at the difference between 320 and 327, it becomes quite complex if one first adds both and then finds the difference. This would be a good problem to give in an interview setting to see how students approached the solution.

Table 4.12

Subtraction Computation

Exercise	Percent Responding		
	Age 9	Age 13	Age 17
Subract 237 from 504.			
Correct	28	73	84
Reversal error	32	5	1
Regrouping errors	4	7	6
Other errors	29	14	8
I don't know or no response	7	1	1

Four multiplication and three division exercises involved computation with whole numbers (see Table 4.13). The low performance of the 9-year-olds is certainly due to their lack of exposure to multiplication and division computation. In fact, the "I don't know" responses on these exercises ranged from 5 to 28 percent. Even when an attempt was made to answer, the errors were more random than those of the older students.

Table 4.13

Results of Multiplication and Division with Whole Numbers

Exercise	Percent Correct		
	Age 9	Age 13	Age 17
Multiplication			
*A. 2-digit by 1-digit, no carrying	65	95	97
*B. 3-digit by 2-digit, no carrying	9	79	86
48 × 4	38	--	--
*2-digit by 1-digit, carrying	34	84	92
671 ×402	3	66	77
Division			
*A. 2-digit divided by 1-digit, fact	74	--	--
*B. 2-digit divided by 1-digit, not a fact	19	--	--
*3-digit divided by 2-digit, no remainder	--	71	85
A. 6)608	--	69	65
B. 28)3052	--	46	50

*Unreleased exercise

By the ages of 13 and 17, performance on simple multiplication exercises was over 80 and 85 percent correct, although performance dropped when three-digit numbers and zeros were involved. Likewise, the performance of 13- and

17-year-olds on simpler division exercises was much higher than when zeros were involved. It is interesting to note that in contrast to all other operations, there was little improvement from age 13 to age 17 on the more difficult division exercises.

Eight of the 34 computation exercises were also given on the first assessment. These exercises, along with the results for both years, are given in Table 4.14. There is little change in the performance on computation between the two assessments, although a slight decline is observable in the performance of 17-year-olds on each exercise.

Table 4.14

Change Data for Whole Number Computation

Exercise	Percent Correct 1972-73			Percent Correct 1977-78		
	Age 9	Age 13	Age 17	Age 9	Age 13	Age 17
What is the sum of 21 and 54?	69	91	--	76	83	--
Add 43 71 75 +92	53	83	94	50	84	92
Subtract 237 from 504.	30	75	89	28	73	84
149 + 327 is how much greater than 149 + 320?	45	74	--	48	75	--
48 × 4	35	--	--	38	--	--
2-digit times 1-digit*	29	84	92	34	84	92
671 ×402	4	69	82	3	66	77
3-digit divided by 2-digit, no remainder*	--	66	87	--	71	85

*Unreleased exercise

Word Problems and Open Sentences

The assessment included one-step and multistep word problems similar to those found in textbooks as well as nonroutine problems. The word problems were presented orally by a tape recording along with the written problem. One strategy that is often used to assist in solving word problems is the writing of an open sentence to represent the problem. The assessment included exercises that required writing open sentences as well as exercises that required only the solving of open sentences. This section includes the results from both types of exercises: word problem and open sentences.

Simple one-step word problems. There were ten exercises (see Table 4.15) that could be classified as simple, one-step word problems; that is, they required one of the four operations, and the answer to the computation was the answer to the problem. The performance on this type of word problem was good if the corresponding computational skills had been developed. This can most clearly be seen for the 9-year-olds. For example, neither subtraction with three-digit numbers nor multiplication of a two-digit number by a one-digit number is a well-developed computational skill, and results on word problems involving these skills reflect this. However, about 80 percent of 9-year-olds responded correctly to addition word problems. Thirteen-year-olds experienced little difficulty with these exercises. The one exception was the larger number division word problem. They did almost as well on it as they did on division computation.

Table 4.15

Simple One-Step Word Problems

Exercise	Percent Correct		
	Age 9	Age 13	Age 17
Addition Exercises			
Paul has 21 stamps in his collection. He buys 54 more from a stamp dealer. How many does he have after he buys them?	82	96	--
Alice collects stamps. She had 45 stamps in her collection. For her birthday her grandmother sent her 6 stamps, her sister gave her 3 stamps, and her best friend gave her 2 stamps. How many stamps does she now have in her collection?	79	--	--
Subtraction Exercises*			
Two-digits, no regrouping (small numbers)	71	--	--
Two-digits, no regrouping (larger numbers)	60	89	--
Two-digits, no regrouping (same numbers as previous exercise except dollars)	61	87	--
Three-digits, regrouping	38	82	--
Multiplication Exercises*			
One-digit times two-digits (small numbers)	28	--	--
One-digit times two-digits (larger numbers)	20	77	--
Division Exercises*			
Division fact	46	--	--
Three-digit divided by two-digit	1	40	57

*Unreleased exercise

Two unreleased subtraction exercises contained the same numbers, but one was an additive situation (How much more?) and the other was a comparative situation that involved money. As can be seen in Table 4.15, there was little difference on these two exercises in the percentage of correct responses for either the 9- or the 13-year-olds.

The results on the addition and subtraction word problems for the 9- and 13-year-olds were similar to those on the previous assessment. However, on the word problem involving the product of a one-digit and a two-digit number, performance fell from 46 percent to 28 percent for the 13-year-olds. Contrast this result to a similar direct computation exercise on which there was a slight improvement from the previous assessment. Is more time being spent now on computation and less on problem solving, or is the sequence in which computation and problem solving are done different now from what it was during the period preceding the first assessment? Either of these could be a plausible explanation for this change in performance.

Complex word problems. Although performance was quite acceptable on simple word problems, it was considerably lower on word problems that required more thought (see Table 4.16). Only one operation was necessary for several of the problems, but there was extra information, missing information, or the answer to the computation was not the answer to the question.

Table 4.16

Complex Word Problems

Exercise	Percent Correct		
	Age 9	Age 13	Age 17
Extraneous Information Exercises			
A. Solves problem (subtraction)*	60	--	--
B. Asks what information is extra	23	--	--
Given 4 numbers in a word problem, answer requires adding 3 of them.*	39	--	--
One rabbit eats 2 pounds of food each week. There are 52 weeks in a year. How much food will 5 rabbits eat in a week?	47	56	--
Missing Information Exercise*			
Identifies missing information, cost problem*	29	--	--
Maria left at noon to take a trip on her bicycle. She rode 5 miles each hour. Later that afternoon Amanda decided to go after her. Amanda rode 10 miles each hour. What else would you need to know in order to find how far the girls rode before Amanda caught Maria?	--	21	37
	--	(24)**	(25)**
Division Exercises			
Divides four-digit by two-digit, need remainder to answer question*	--	19	--
A man has 1310 baseballs to pack in boxes that hold 24 baseballs each. How many baseballs will be left over after the man has filled as many boxes as he can?	--	29	--

*Unreleased exercise
**Additional correct percentage

There were three exercises that contained one extra number in the word problem. In the first exercise described in Table 4.16, 60 percent of the 9-year-olds could solve the problem, but only 23 percent could identify the

26

piece of extra information. This, along with the 18 percent who responded "I don't know," illustrates that students are not accustomed to being asked such a question. Another unreleased exercise gave four items and their prices on a menu and asked the total cost of three of them. Thirty-nine percent of the 9-year-olds responded correctly, but almost the same number added all four items. The rabbit exercise showed that the 13-year-olds also have difficulty with extraneous information. Almost one-fourth of the 13-year-olds found the product of all three numbers instead of disregarding the extra number.

Each age group was given one exercise that had missing information. A little less than one-third of the 9-year-olds could identify what additional information was needed in order to solve an unreleased open-ended exercise that asked for the total cost of a number of items. The older two groups were administered the problem with missing information found in Table 4.16. In addition to the 21 and 37 percent of the 13- and 17-year-olds who responded that you would need to know what time Amanda left, about one-fourth of each age group responded that you would need to know how long Maria or Amanda rode. If this latter response was interpreted as how long either Maria or Amanda rode before Amanda caught up, then it could be considered as the necessary missing information. With this interpretation, about half of each of the older groups identified the missing information.

The two division exercises in Table 4.16 probably best illustrate the lack of thought that many students give to problem solving. The responses to the released exercise, illustrative of both exercises, are found in Table 4.17. At least two-thirds of the 13-year-olds knew that the problems called for division, but only about one-fourth realized that the remainder and not the quotient was the answer. Indeed, for this problem the quotient was an unreasonable answer, as was expressing the remainder as a fraction. Both of these exercises indicate that more emphasis needs to be placed on remainders and their meanings and interpretations.

Table 4.17

Responses to Remainder Exercise

Exercise	Percent Responding Age 13
A man has 1310 baseballs to pack in boxes which hold 24 balls each. How many baseballs will be left over after the man has filled as many boxes as he can?	
Correct	29
Other	19
Divided, did not tell how many left over	22
Divided, wrong whole remainder	7
No response	5
I don't know	9

Open sentences. There were two types of tasks involving open sentences: writing or identifying an open sentence that represents a given problem and solving open sentences.

The three exercises reported in Table 4.18 are translation problems. Both 9- and 13-year-olds were more successful at choosing than writing an open sentence that represents a word problem. This, of course, is partially due to the multiple-choice format of the exercises that require choosing an open sentence. In these examples, the difference in performance can also be partly accounted for by the use of the word equation in the writing exercise.

Table 4.18

Translating Word Problems to Open Sentences

Exercise	Percent Correct		
	Age 9	Age 13	Age 17
Choosing addition sentence exercise			
A. Paul has 21 stamps in his collection. He buys 54 more. Which number sentence tells how many he has after he buys them?	67	92	96
B. Mike took 15 cents to the store. He saw a pen that costs 23 cents. Which number sentence tells you how much more he needs to buy the pen?	50	84	90
Writing sentence exercise			
Janelle had 173 matchbook covers in her collection. Her aunt sent her some more. She now has 241 matchbook covers.			
A. Write an equation to show how many matchbook covers Janelle's aunt sent her.	--	13	43
B. Solve this equation.	--	63	82
Choosing division sentence exercise*			
Chooses division sentence to represent			
A. Partioning word problem	--	46	82
B. Measurement word problem	--	44	82

*Unreleased exercise

Overall, about 80 percent of the 13-year-olds chose the correct sentence in all exercises. The performance for 9-year-olds depended more on the type of operation involved. The students were asked to write an equation about Janelle's matchbook covers (Table 4.18) and to solve the equation. Note that more students <u>solved</u> than wrote the equation. This percentage actually reflects those who solved the word problem with or without the equation. The contrast between the results of solving a problem and writing an equation should raise questions about when and how the equations or open sentences help with finding the solution to a problem. As can be seen in what follows, many students were able to solve open sentences given to them. They seem to lack the ability to tie the two skills together.

If open sentences (or equations) are to be used to represent word problems, then students must be able to solve open sentences. The exercises in Table 4.19 required students to solve the open sentence given without a story problem.

While certainly not every type of open sentence was administered, several patterns are observable from the set. The size of the numbers in the sentence influenced the level of performance. Students were able to handle the open sentence when only basic facts were involved. Their level of performance on the other open sentence partly depended on their level of corresponding computation.

There was considerable difference in 13-year-olds' ability to handle a sentence with a ☐ instead of a variable, as can be noticed in the missing factor exercises. Ninety-one percent could identify the missing factor of a multiplication fact when a ☐ was used, compared to 65 percent when <u>n</u> was used in the sentence.

Table 4.19

Solving Open Sentences

Exercise	Percent Correct	
	Age 9	Age 13
Missing addend exercises*		
A. Solve for missing addend, addition fact sentence	93	--
B. Solve for missing addend, subtraction sentence	74	--
A. Solve for missing addend, 2-digit numbers, no regrouping	63	89
B. Solve for missing addend, 3-digit numbers, no regrouping	41	80
Subtraction open sentence exercise*		
☐ - 19 = 32; what number should go in the ☐ to make the sentence TRUE?	15	54
Missing factor exercises*		
Solve for missing factor; multiplication fact	66	91
Solve for missing factor, divides 3-digit by 1-digit	--	57
A. Solve for missing factor; uses n in sentence, multiplication fact	--	65
B. Solve for missing factor, uses n in sentence, 3-digit divided by 2-digit	--	30

*Unreleased exercise

The only exercise in which the ☐ was first in the sentence was the open subtraction sentence, ☐ — 19 = 32 (Table 4.19). This proved to be much more difficult than any of the other open addition or subtraction sentences for both 9- and 13-year-olds. About one-fourth at each age subtracted 19 from 32.

Multistep and nonroutine word problems. There were only four multistep word problems involving whole numbers, that is, problems that required two operations. Performance was lower on these than on the simple one-step problems but about the same as on the more complex one-step problems. None of the exercises were released, and so two illustrative exercises will be described.

Both the 9- and 13-year-olds were given an exercise that asked the amount of change after a purchase of two items. Thirty-seven percent of the 9-year-olds chose the correct answer, and another 28 percent chose an answer that indicated they knew what to do but made an error in adding or subtracting. Almost three-fourths of the 13-year-olds could do this problem, but only about one-fifth of them could solve a problem that required the following two steps: subtract a five-digit number from a five-digit number and divide the three-digit difference by a two-digit number. About one-fifth of this age group stopped after subtracting, another 11 percent added the two larger numbers, and 7 percent divided only. Two-thirds of the 17-year-olds correctly solved this multistep problem.

Several nonroutine problems were administered. These were problems unlike those generally found in textbooks and required the application of knowledge, skills, and understanding to somewhat unfamiliar situations. These problems were quite varied, and success depended on the nature of the problem. The results of one released exercise are presented in Table 4.20. Very few 9-year-olds and only about one-fourth of the 13-year-olds were successful.

As can be seen from the incorrect responses, many 9-year-olds added the three numbers listed in the problem. This was a strategy that this age group often followed when unsure of the method of solution. Most responses were nonclassified, indicating a lot of random solutions.

Table 4.20

Nonroutine Exercise: 9- and 13-Year-Olds

Mike's building set has

    60 long pieces

    60 short pieces

and 60 nuts with bolts

How many of these
can he make?

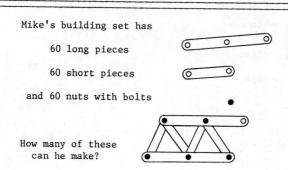

	Percent Responding	
	Age 9	Age 13
Correct (12)	3	24
15	1	7
30	5	9
10	5	9
60	4	3
180	20	9
Other	38	31
I don't know or no response	23	8

The 17-year-olds were given the problem in Table 4.21. Forty-two percent were successful in selecting the correct response from the three choices. As with all the nonroutine problems, direct observation of how students solved these would give more information than was possible to gather on this assessment.

Summary

    As mentioned in the introduction to this section, there was a great emphasis on whole numbers in the assessment. Because of the great number of exercises, the results are only highlighted here by age levels.
    The 9-year-olds were successful on counting, writing numbers from oral words, recalling basic addition and subtraction facts, computing simple addition and subtraction problems, and solving simple addition and subtraction word problems and open sentences. The 13-year-olds performed much better than the 9-year-olds on all tasks and were also successful in problems involving simple multiplication and division. While 17-year-olds performed better than the 13-year-olds on every exercise, there was not much difference between the two age levels on the simple problems. Greater difference was found on harder exercises.
    None of the age groups were impressive in their performance on any of the

Table 4.21

Nonroutine Exercise: 17-Year-Olds

Juan's mother has three five-dollar rolls of dimes and two ten-dollar rolls of quarters to use for Juan's school lunches.

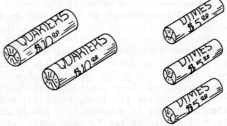

If Juan takes exactly forty-five cents to school every day for his lunch, which of the following statements is true?

	Percent Responding
⬭ He uses all of the quarters before all of the dimes.	17
⬤ He uses all of the dimes before all of the quarters.	42
⬭ He spends all of both coins at the same time.	36
⬭ I don't know	5

word problems other than the simple one-step ones. Division computation was the other most difficult task for the older two groups.

The difference in performance between the two assessments on exercises involving whole numbers was not great. Several examples of change were given throughout the section; these are more meaningful than a summary statement.

Fractions

Overview of Results

About forty exercises involved only fractions. About half of these dealt with fraction concepts, language, or equivalency; the rest were almost evenly split between pure computation and word problems. The major findings were the following:

1. Concepts and models underlying fractions are not well developed by age 9. Although older students do not have much difficulty relating fractions to pictorial models, they do not seem to realize that these models may be helpful in other situations.
2. Computational skills with fractions are not well developed by age 13 or age 17. The skills that have been mastered appear to have been done with little understanding.
3. Results for 13- and 17-year-olds on word problems involving multiplication of fractions were substantially lower than on corresponding computation exercises.

Fraction Concepts

Several basic ideas about fractions were assessed, including the language of fractions, pictorial models, order, and equivalency. Results from all of these exercises give some indication of the students' ability to handle fraction concepts.

Exercises for 9-year-olds that dealt with the language associated with fractions required them to either choose or write the symbol representing the fraction when given the verbal description of it. About seven-eighths chose the correct symbol, but only a little over one-half wrote the correct symbol for the same fraction. When students could write the symbol, they could do so for any proper fraction and not just familiar ones, such as two-thirds.

The other language exercises required 13-year-olds to identify such terms as <u>denominator</u>, <u>improper fraction</u>, or <u>mixed numeral</u>. About three-fourths of the 13-year-olds were successful on these knowledge exercises.

Performance on several exercises involving pictorial models showed that the concept of a fraction as a part of a whole is a developing concept for 9-year-olds and is fairly well developed for 13-year-olds. When 9-year-olds were given a region divided into the same number of parts as the denominator of the fraction, 62 percent could shade the region to show the fraction. When the region was divided into more parts than the number in the denominator, however, they experienced much difficulty. One such exercise is shown in Table 4.22. Although the phrase <u>fractional part</u> may have accounted for some of the errors, another exercise that did not use the phrase produced similar results.

Table 4.22

Fractional Part of a Region

What fractional part of the figure is shaded?

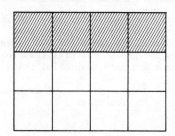

Responses	Percent Responding	
	Age 9	Age 13
Acceptable responses $\frac{1}{3}$, $\frac{4}{12}$, .33	20	82
Unacceptable responses $\frac{1}{4}$, .25	5	4
Top 4, top part, $\frac{4}{8}$	36	6
Other	15	6
I don't know	17	1
No response	7	1

One exercise asked both 9- and 13-year-olds to write a fraction that rep-

resented a set model. The results are given in Table 4.23. The 9-year-olds
focused on the number of colored squares and not on the fractional part. Although performance was lower on this exercise than on those involving regions,
it was slightly higher for both ages than on similar exercises given during
the first assessment.

Table 4.23

Fractional Part of a Set

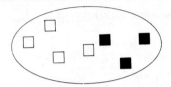

The above set is made up of green and white squares. What
fractional part of the squares is green?*

| Responses | Percent Responding | |
	Age 9	Age 13
Acceptable response		
$\frac{3}{8}$	12	72
Unacceptable responses		
$\frac{5}{8}$	1	2
3	58	8
$\frac{3}{5}$ or $\frac{5}{3}$	4	11
"right" part	3	0
Other	14	7
I don't know	5	0

*This exercise is similar to an unreleased exercise.

Thirteen-year-olds have not mastered number line representations for fractions. Slightly more than half of them could locate 1/2, and less than half
could locate 1 3/4, on the appropriately marked number lines shown in Table
4.24. Note that almost 20 percent marked 1/2 at a location midway between 1
and 2. Another 14 percent marked 1 or the halfway mark between the labeled
endpoints. The second part of this exercise illustrates the same type of error
often found in measuring exercises--that of not focusing on the unit.

Several other exercises that did not involve pictures give additional insight into the students' understanding of fractions. Nine-year-olds were
asked how many parts (halves, fourths, etc.) made a whole. About half of them
chose the correct number. About half of the 13-year-olds could relate a fraction to the division interpretation. About one-third of the 17-year-olds realized what whole number an infinite sum of decreasing fractions would approach.

Very few students at any age ordered the six fractions shown in Table
4.25. In another exercise, however, over half of the 13-year-olds and four-
fifths of the 17-year-olds chose a fraction between two others. These two exercises do not give enough information to make any generalizations about students' ability to order fractions.

Table 4.24

Number Line Representation

Exercise	Percent Responding Age 13
A. Mark an X on the number line where $\frac{1}{2}$ should be.	

$$\xleftarrow{\hspace{1cm}} \underset{0}{\quad} \underset{1}{\hspace{3cm}} \underset{2}{\hspace{3cm}} \xrightarrow{\hspace{1cm}}$$

Acceptable response, any mark between $\frac{1}{4}$ and $\frac{3}{4}$	58
Unacceptable responses:	
Any mark between $1\frac{1}{4}$ and $1\frac{3}{4}$	21
A mark at 1	14

Exercise	
B. Mark an X on the number line where $1\frac{3}{4}$ should be.	

$$\xleftarrow{\hspace{1cm}} \underset{0}{\quad} \underset{1}{\hspace{2cm}} \underset{2}{\hspace{2cm}} \underset{3}{\hspace{2cm}} \xrightarrow{\hspace{1cm}}$$

Acceptable response, any mark between $1\frac{5}{8}$ and $1\frac{7}{8}$	43
Unacceptable responses:	
A mark at $1\frac{3}{8}$	20
A mark at $1\frac{4}{8}$ or any mark between $1\frac{3}{8}$ and $1\frac{5}{8}$	9
A mark at $1\frac{5}{8}$	6

Table 4.25

Ordering Fractions

Exercise	Percent Correct		
	Age 9	Age 13	Age 17
Arrange the following fractions from the SMALLEST to the LARGEST $\frac{5}{8}, \frac{3}{10}, \frac{3}{5}, \frac{1}{4}, \frac{2}{3}, \frac{1}{2}$	0	2	12
Smallest Fraction ___ ___ ___ ___ ___ ___ Largest Fraction			

A major portion of the exercises on fractions required the recognition of varied representations for the same quantity. While three-fourths of the 13-year-olds and over nine-tenths of the 17-year-olds could reduce 3/9 to lowest terms, performance dropped considerably when the fractions were less familiar. Results in Table 4.26 show some of the responses for more difficult fraction exercises. Levels of performance similar to those shown in Table 4.26 were obtained on exercises requiring equivalency between mixed numerals and improper fractions.

Table 4.26

Fractions in Lowest Terms

Exercise	Percent Responding	
	Age 13	Age 17
Reduce the following fractions to lowest terms: $\frac{14}{35} =$		
Correct response	57	78
Unacceptable responses		
$\frac{14}{35}$	2	2
$\frac{2}{7}$	3	3
$\frac{7}{5}$	4	3
Other unacceptable responses	18	9
I don't know	4	2
No response	11	3

Computation with Fractions

This assessment included computation exercises requiring students to add, subtract, or multiply. No exercises involving division were administered, and the majority of the computation exercises were given only to 13- and 17-year-olds.

The 9-year-olds were given two exercises in which they had to add fractions. One of the exercises included a picture of a circular region divided into parts that corresponded to the denominator of the two like fractions to be added. While only 11 percent responded correctly, 16 percent responded correctly when given the same problem without a picture. Less than 2 percent correctly added fractions with unlike denominators. Nine-year-olds have had little, if any, experience adding fractions, so these results are not unexpected.

The results in Table 4.27 demonstrate that about two-thirds of the 13-year-olds and five-sixths of the 17-year-olds could add two fractions with like denominators. Performance dropped when the denominators were unlike; however, the complexity of the unlike denominators had little to do with the level of performance. Apparently, if students have learned a computational algorithm, they can apply it in most situations. If they have not mastered an algorithm, then they cannot reconstruct it or fall back on intuitive models to solve simpler problems.

35

Table 4.27
Fraction Addition Exercises

Exercise	Percent Correct Age 13	Age 17
A. $\frac{4}{12} + \frac{3}{12} =$	74	90
B. $2\frac{3}{5}$ $+4\frac{4}{5}$	63	77
C. $\frac{3}{4} + \frac{1}{2} =$	35	67
D. $\frac{1}{2} + \frac{1}{3} =$	33	66
E. $2\frac{2}{5} + 5$	43	65
F. $4\frac{1}{4}$ $+3\frac{2}{5}$	46	60
G. $\frac{7}{15}$ $+\frac{4}{9}$	39	54

The conclusion that many students have only a rote procedure for performing addition of fractions is further substantiated by the estimation exercise in Table 4.28. Lower performance on the computation exercises indicates that many students who can successfully apply an algorithm for adding fractions have little understanding of the underlying concepts and processes.

Table 4.28

Results of Estimating a Fraction Sum

ESTIMATE the answer to $\frac{12}{13} + \frac{7}{8}$. You will not have time to solve the problem using paper and pencil.

Responses	Percent Responding Age 13	Age 17
○ 1	7	8
● 2	24	37
○ 19	28	21
○ 21	27	15
○ I don't know	14	16

One exercise administered to 13- and 17-year-olds asked for the least common denominator of two fractions (see Table 4.29). Note that these are the same fractions as in part G in the computation table, Table 4.27. Fewer students could find the least common denominator or even a common denominator than could add the two fractions. This could be a language problem, but it is also an indication of the lack of understanding that a common denominator must be found when adding fractions.

Table 4.29

Results of Least Common Denominator Exercise

Exercise	Percent Responding	
	Age 13	Age 17
Find the least common denominator of $\frac{7}{15}$ and $\frac{4}{9}$		
Acceptable response 45	24	42
Unacceptable responses 135 or 15 × 9	4	6
3 or a "least" common factor	16	20
Other	30	23
I don't know	17	4
No response	9	5

There were fewer subtraction than addition exercises involving fractions, but the results followed the same pattern. As one might expect, the results on subtraction tended to be lower than on addition. Computations requiring regrouping before subtracting were especially difficult. In an unreleased exercise, only about one-fourth of the 13-year-olds and less than one-third of the 17-year-olds chose the appropriate regrouping (e.g., 2 4/7 = 1 11/7).

No formal multiplication computation exercises were administered to 9-year-olds, but they were asked to find a fractional part of a whole number, such as 1/3 of 9. About one-fourth of the 9-year-olds were successful on such exercises.

About one-half of the 13-year-olds and two-thirds of the 17-year-olds could multiply two proper fractions or a whole number by a proper fraction. Performance dropped considerably when at least one of the factors was a mixed number. The results of one exercise in Table 4.30 illustrate the level of performance for the mixed number multiplication. The most common error was multiplying the whole numbers and fractions separately, with about 28 percent of both age groups making this error.

Table 4.30

Multiplication of Fractions and Mixed Numbers

Exercise	Percent Correct	
	Age 13	Age 17
Find the products:		
$\frac{3}{4} \times \frac{3}{5}$	69	74
$3 \frac{1}{2} \times 6 \frac{2}{3}$	28	42
$\frac{1}{6} \times 7 \frac{1}{3}$	30	44

Word Problems Involving Fractions

Ten exercises involved fractions in one-step or multistep word problems. These exercises were administered to the older groups, except for one exercise involving addition. Only about 13 percent of the 9-year-olds were successful on this exercise, and so the discussion here will center on the other age groups.

The performance on all four of the word problems involving the addition or subtraction of fractions was similar to the results of the exercise presented in Table 4.31. Few differences in the percentages of correct responses were due to factors such as whether the problem required addition or subtraction, had a picture with it or not, or required choosing a phrase to represent the problem or required solving the problem. About half of the 13-year-olds and three-fourths of the 17-year-olds could do these tasks with numbers involving halves, fourths, and eighths. When we compare these results with the computation results, it appears that the word problem may have assisted in the computation. Results are higher on the word problems than on computation exercises involving the same type of fractions.

Table 4.31

Word Problem Involving Addition of Fractions

To make each cake you need 2 1/4 cups of sugar for the frosting and 1 1/2 cups of sugar for the cake. How many cups of sugar are needed for the cake and frosting together?

	Percent Responding	
	Age 13	Age 17
○ $3\frac{2}{6}$ cups	31	10
○ $3\frac{2}{4}$ cups	10	7
● $3\frac{3}{4}$ cups	54	81
○ $3\frac{2}{2}$ cups	3	1
○ I don't know	1	1

There were two unreleased exercises, each of which involved the multiplication of the same two proper fractions. Performance on each was consistent even though the context varied; only 17 percent and 30 percent of the 13- and 17-year-olds found the correct product. These percentages were much lower than the results on corresponding computation exercises. This again points to the conclusion that some computation algorithms may be learned rotely with little ability to apply them even to simple situations.

Just as on multistep word problems with whole numbers, the performance levels on multistep word problems with fractions were low. The only released exercise, presented in Table 4.32, can be considered a simple two-step problem. The 17-year-olds had difficulty with this exercise; only about 30 percent could take one-half of 3 3/4. Another 22 percent gave a reasonable answer, indicating that they knew to take half but did so incorrectly. Performance on the other exercises varied according to the complexity of the problem or the complexity of the fractions.

Table 4.32

Two-Step Word Problem with Fractions

A recipe for punch calls for 3 3/4 cups of pineapple juice for 10 people. How much pineapple juice could be used to make the same punch to serve five people?

	Percent Responding Age 17
Correct (1 $\frac{7}{8}$ cups)	13
1 $\frac{7}{8}$ with no unit or incorrect unit	16
Incorrect answer between 1 $\frac{1}{2}$ and 2	22

Summary

The assessment results certainly reiterate what has been known about students' ability to work with fractions: they have difficulty with many of the concepts, and their computational work often appears to be a nonmeaningful routine.

Decimals

Overview of Results

As with the other types of number exercises, the approximately forty-five decimal exercises on this assessment involved concepts and properties, computations, and applications. Most of the application exercises dealt with money and are thus only tangentially related to decimals. The majority of the exercises were administered only to the 13- and 17-year-olds. The major results on the decimal exercises are the following:

1. Nine-year-olds demonstrated little familiarity with decimals and treated them as whole numbers.
2. Although 13- and 17-year-olds appeared to have facility with tenths and hundredths, their competency was less developed for thousandths and smaller decimal numbers.
3. The ability to relate decimals and common fractions was not well established even by age 17.
4. Performance by 13- and 17-year-olds on computation exercises was at about the same level as their knowledge of basic decimal concepts, that is, about 50 percent correct for 13-year-olds and 70 percent correct for 17-year-olds. Performance was lower on division exercises that involved a decimal divisor.
5. Applications with money varied according to the type of problem. Nine-year-olds seemed to have an easier time with problems presented in a money context than with corresponding decimal computation.

Decimal Concepts and Properties

In developing decimal concepts, it is necessary to be able to relate symbols to words. The results in Table 4.33 represent the easiest such exercise administered on this assessment. Most 9-year-olds have not studied decimals, and they responded by ignoring the decimal and treating the number as a whole number. The older two groups were able to relate their knowledge of decimals to this exercise and performed much better than the 9-year-olds, but results on a second exercise (see Table 4.34) reveal that they do not have complete mastery of this skill. Both 13- and 17-year-olds had more difficulty with the second part of this exercise than they did with the exercise in Table 4.33. Although this may have been because of the change in the order of the stem and response, it may have resulted from coming immediately after a more difficult part. Part A, along with other exercises involving thousandths, produced

Table 4.33

From Symbols to Words

What is the number in the box?*

```
7.94
```

	Percent Responding		
	Age 9	Age 13	Age 17
○ Seventy-nine and four tenths	12	5	2
○ Seven hundred ninety-four	59	13	5
● Seven and ninety-four hundredths	21	79	92
○ Seventy-nine hundredths	2	2	1
○ I don't know	5	1	0

*This exercise is similar to an unreleased exercise.

Table 4.34

From Words to Symbols

Exercise	Percent Responding		
	Age 9	Age 13	Age 17
A. Which of the following is thirty-seven thousandths?			
● 0.037	3	54	72
○ 0.37	2	3	4
○ 37	2	2	2
○ 37000	88	40	22
○ I don't know	5	0	0
B. Which of the following is one and twenty-four hundredths?			
○ 1.024	11	14	15
● 1.24	19	65	76
○ 2401	6	1	0
○ 12400	57	18	7
○ I don't know	6	1	0

lower results than exercises with tenths or hundredths. Note that the 13-year-olds, like the 9-year-olds, ignored the decimal point and treated the number as a whole number. It may be hypothesized that the zero in the answer is partly responsible for the lower result, because in all such exercises involving a zero, the performance was slightly lower. The older groups, however, were able to handle a zero with hundredths at much nearer the level they could handle other hundredths than was true in the case of thousandths.

Several exercises required students to relate decimals to a number line. The results of two of these exercises are given in Table 4.35. While 13- and 17-year-olds could choose the decimal that named a clearly identifiable point, they had difficulty approximating a decimal that represented a position between two clearly identifiable numbers. This relates closely to their ability to give a decimal between any two decimals. About 40 and 75 percent, respectively, of the 13- and 17-year-olds could choose a decimal between two given ones, such as .35 and .36. Performance on such tasks dropped about 25 percentage points if they were asked to give (rather than to choose) the decimal, and if the decimal involved thousandths.

Table 4.35

Number Line Exercises: Representation of Decimals

Exercises	Percent Responding		
	Age 9	Age 13	Age 17

A. What is the number at point A?

	Age 9	Age 13	Age 17
○ .11	11	5	3
● 1.1	46	80	87
○ 1.5	14	8	6
○ 11	7	2	0
○ I don't know	16	3	2

B. The arrow shows a point on the number line. Write a number that could represent that point.

	Age 9	Age 13	Age 17
Correct response, any number from 3.26 to 3.29		34	64
Unacceptable responses			
Any number from 3.25 to 3.259 or 3.291 to 3.299		10	8
Any number from 3.20 to 3.249		13	7
3.3		9	3
Other		27	12
I don't know		7	6

One way to introduce decimals is through common fractions, and several exercises examined the students' ability to relate the two notations. The most basic relationship involves common fractions in tenths, hundredths, and so forth, and the equivalent decimal fractions. It was evident from the one exercise given to 9-year-olds that they had not developed any connection between

fractions and decimals. Fewer than 1 percent could give the decimal equivalent of a common fraction expressed in tenths.

Since 9-year-olds are not familiar with decimals, most of the exercises
examining the relationship between common fractions and decimals were given
to the older two groups. About 60 percent of the 13-year-olds and 75 percent
of the 17-year-olds could express common fractions, tenths and hundredths, as
decimals. Performance dropped to about 40 and 60 percent when the common
fraction was an improper fraction in the hundredths. Likewise, when thousandths
were involved, performance was below 50 percent for both age groups.

Performance on exercises that required changing a decimal to a fraction
decreased slightly from tenths to thousandths. Approximately 50 and 70 percent of the 13- and 17-year-olds wrote the equivalent fraction for decimals
in the tenths and hundredths and about 40 and 60 percent wrote the equivalent
fractions for decimals in the thousandths.

A more difficult task is changing a fraction not expressed in tenths,
hundredths, and so forth, to a decimal. Since 13-year-olds did not have a
firm grasp of the fact that a fraction can express a division (see "Fractions"), it is not surprising that they had difficulty with converting fractions to decimals. The results of an unreleased exercise similar to the
questions shown in Table 4.36 reveal this difficulty for both of the older
groups. In both parts, the most common errors involved using in some way one
or both of the digits in the fraction. For example, more 13-year-olds chose
.85 as the equivalent of 5/8 than chose the correct answer.

Table 4.36

Fraction-Decimal Equivalency

| Exercise | Percent Responding | |
	Age 13	Age 17
A. Which decimal is equal to $\frac{1}{5}$?*		
○ .15	12	5
● .2	38	75
○ .5	38	18
○ .51	3	1
○ I don't know	6	1
B. Which decimal is equal to $\frac{5}{8}$?*		
○ .6	7	5
● .625	27	59
○ $.\overline{714285}$	3	4
○ $.8\overline{5}$	30	11
○ I don't know	30	18

*This exercise is similar to an unreleased exercise.

The ability to order decimals is another indication of students' compe-
tence with decimals. The results of the released exercise in Table 4.37 show
great variation in perfomance on the different parts of the exercise. In
part A, most 13- and 17-year-olds realized that a number greater than 1 is
more than a number less than 1, but in part C as many 13-year-olds chose the
greatest number--by ignoring the decimal and treating the numbers as whole
numbers--as chose the correct response. Only about 10 and 25 percent, re-

spectively, of the 13- and 17-year-olds were able to order decomals from least to greatest when given five decimals similar to those in part C.

Table 4.37

Ordering Decimals

Exercise	Percent Responding	
	Age 13	Age 17
A. Which number is greater?		
● 1.9	81	94
○ 0.23	18	5
○ I don't know	1	0
B. Which number is greater?		
○ 1.15	20	23
● 1.36	79	76
○ I don't know	1	0
C. Which number is the greatest?		
○ .19	2	2
○ .036	4	3
○ .195	47	21
● .2	46	72
○ I don't know	1	2

Although it is clear that there is variation in performance on the types of exercises discussed here, the average performance for 13-year-olds and 17-year-olds was approximately 50 percent and 70 percent, respectively.

Decimal Computation

The only decimal computation exercises given to the 9-year-olds were addition exercises. A representative addition exercise is given in Table 4.38. The results show that 9-year-olds ignored the decimal point and computed as with whole numbers. Given the same addition exercise expressed in terms of money, over one-third of the 9-year-olds correctly computed the sum. In general, 9-year-olds showed little ability to handle addition with decimals. Although performance on the addition exercise in Table 4.38 was better for the older students than for the 9-year-olds, it is important to note that one-fourth of the 13-year-olds and one-eighth of the 17-year-olds placed the decimal point to the left of the sum (.130 or .13), just as it appears in the addends (.70, .40, .20).

On a wide variety of addition and subtraction exercises, the performance for 13-year-olds ranged from 34 to 87 percent correct and for 17-year-olds from 63 to 93 percent correct. The lowest performance levels were on exercises involving horizontal notation and decimals expressed in different units (e.g., subtracting a decimal expressed in hundredths from one expressed in tenths). There was some indication from the results reported in Table 4.39 that students recognized a need to "line up the decimal points." However, when given an exercise in horizontal format, they either failed to do so or they did not know how to handle the unlike units.

43

Table 4.38

Decimal Addition

Exercise	Percent Responding		
	Age 9	Age 13	Age 17
Add the following .70 + .40 + .20			
Acceptable responses 1.3, 1.30	11	58	79
Unacceptable responses .13 or .130	4	23	12
130	44	6	4
13 with other decimal placements	13	6	1
Other	24	6	6
I don't know	4	1	0

Table 4.39

Analyzing Errors in Computation

Exercise	Percent Responding	
	Age 13	Age 17
Liz wrote this addition problem on her paper:		

$$6.32$$
$$4.3$$
$$.113$$
$$\overline{.788}$$

Her teacher marked it wrong. Which one of the following best explains why it was marked wrong?		
⚪ Liz does not know her basic addition facts.	2	1
⬤ Liz does not understand decimal places.	85	95
⚪ Liz does not know the rule for "carrying" in addition.	3	1
⚪ The teacher made a mistake; Liz's answer is correct.	10	3
⚪ I don't know	1	0

Approximately one-half of the 13-year-olds and three-fourths of the 17-year-olds correctly multiplied decimals or chose the correct placement of decimal points in a given product. Depending on the difficulty of the multiplication problem, the errors in multiplying decimals were about evenly split between a multiplication error and placement of the decimal. In general, about 12 percent of the 13-year-olds and 8 percent of the 17-year-olds who multiplied correctly made an error in placing the decimal. For these students as well as for those who incorrectly answered the exercise, one can conjecture that they are not in the habit of looking at the reasonableness of their answers.

The division exercises with decimals involved simple divison with small divisors. Seventy-six percent of the 13-year-olds and 90 percent of the 17-year-olds correctly divided a decimal by a whole number, such as 8.4 ÷ 4. Performance dropped to 55 percent when the divisor was tenths, such as 8.4 ÷ .4, and to 37 percent when the divisor was hundredths, such as 8.4 ÷ .04. These examples, along with another unreleased exercise that asked students to identify the way the exercise could look after the "first step" in dividing by a decimal, show that division by a decimal is not a well-developed skill.

Estimation exercises involving computation with decimals give additional information about 13- and 17-year-olds' skills with decimals. An unreleased addition exercise had three parts in which students had to choose an estimate of the sum of two or three decimals. The part that gave students the most difficulty was similar to the exercise shown in Table 4.40. Students did not realize that the estimate depended on the decimal expressed in hundredths. The results of the other parts of the addition estimation exercise were about 12 percentage points higher than this part.

The size of the numbers greatly influenced the level of performance on the estimation of decimal multiplication. Although performance was low on the multiplication exercise in Table 4.40, it was about the same for 13-year-olds and about 15 percentage points higher for 17-year-olds when the two factors were between 1 and 10. Performance was much higher (62 and 78 percent) for a two-digit number multiplied by a decimal less than 1.

The results of a released exercise involving division computation are also given in Table 4.40. Performance was low on all the parts of the exercise but varied according to the type of problem. The easiest part was one in which the given estimate was the exact answer and that was also fairly easy to compute directly. It appears that either the skill of rounding numbers in order to make an estimate is not well developed or students are not accustomed to being asked for an estimate only.

Decimal Applications

As mentioned in the introduction, almost every word problem that involved decimals was given in the context of money. Although these are important applications, the assessment did not provide much information about the use of decimals in other types of applications. The exercises reported here, along with the applications of percents, could be considered consumer applications.

There were six exercises typical of the one-step problems often found in textbooks. The computation needed for several of these was also given in the decimal computation exercises. For example, the exercise reported in Table 4.41 required adding $.70, $.40, and $.20. On the same exercise given in a pure computation format, 9-year-olds scored about 22 percentage points lower; they scored about 46 percentage points lower when neither money nor a story problem was involved. About 40 percent of the 9-year-olds could figure out the difference between two amounts, both of which were less than $6.00. Although there was no parallel exercise, there was a similar addition problem with decimals in the hundredths. Only 25 percent of the 9-year-olds were successful on this exercise. Together, these exercises give some indication that the money context of a story problem does assist this age group in finding the solution.

Table 4.40
Estimating Decimal Computations

ESTIMATE the answer to these problems. You will not be given enough time to calculate the answer by using paper and pencil. Fill in the oval that is the CLOSEST to your ESTIMATE.**

	Percent Responding	
	Age 13	Age 17
Addition Exercise*		
.02 + .0002 + .000008		
○ 2	5	5
● .02	12	33
○ .00022	68	52
○ .2	5	5
○ I don't know	10	5
Multiplication Exercise*		
.3837 × .22		
○ .008	47	41
● .08	16	25
○ .8	9	12
○ 8	12	10
○ I don't know	15	11
Division Exercise		
A. 30 ÷ 317		
● .1	7	15
○ .01	22	30
○ 1	6	8
○ 10	59	41
○ I don't know	5	6
B. 250 ÷ .5		
○ 50	61	47
● 500	25	39
○ 1000	3	3
○ 1250	4	6
○ I don't know	6	4
C. .239 ÷ .4		
○ .006	24	20
○ .06	34	32
● .6	19	28
○ 6	8	7
○ I don't know	13	12

*This exercise is similar to an unreleased exercise.
**Instructions are paraphrased here to fit all the exercises.

Table 4.41

Addition Application

Exercise	Percent Responding		
	Age 9	Age 13	Age 17

MENU			
Hamburger	.85	Milk	.20
Hot Dog	.70	Soft Drink	.15
Grilled Cheese Sandwich	.55	Milk Shake	.45
French Fries	.40	Ice Cream	.40

Sue had a hot dog, french fries, and milk.
How much did she spend?

	Age 9	Age 13	Age 17
○ $1.20	9	2	1
● $1.30	57	92	95
○ $1.40	9	2	2
○ $1.50	16	3	1
○ I don't know	6	0	0

Thirteen and 17-year-olds were given a word problem that required the multiplication of a two-digit number by a decimal. Performance on this exercise was at the 42 and 76 percent level for 13- and 17-year-olds respectively, or about 30 and 10 percentage points lower than on the corresponding computation exercise. Of students who knew that the word problem required multiplication, many more made a mistake in placing the decimal point than in doing the computation. Evidently not even the context of the word problem helped these students decide if their answer was reasonable.

The exercise in Table 4.42 involved only one computation if the table was read carefully. There was a wide variety of uninterpretable answers (32 percent), a large percentage of "I don't know" or no response (29 percent), and a large number (19 percent) who divided 606 by 9.09. In all, it appeared that about 16 percent of the 17-year-olds knew to divide 9.09 by 606; out of these,

Table 4.42

Application Problem

A bill for electricity contains the following information:

Present Reading	Previous Reading	Consumed	Bill
1548 kw-hr	942 kw-hr	606 kw-hr	$9.09

How much is the customer paying per kilowatt hour for electricity?

	Percent Responding Age 17
Correct answer	5
Correct numerical answer, wrong or no units	4
Attempted correct division, wrong answer	7
Divided 606 by 9.09	19
Other unclassified errors	32
Wrong operation	4
No response or I don't know	29

about 5 percent wrote the correct answer, another 4 percent used the wrong unit or no unit with the correct numbers, and 7 percent made an error in the division.

There were six unreleased exercises involving money in multistep problems given to 13- and 17-year-olds. Results varied according to the type of problem. The easiest asked for the change from a bill for the purchase of two items, each costing less than $10.00. Sixty-two percent of the 13-year-olds and 77 percent of the 17-year-olds gave the appropriate amount of change. The most difficult was one similar to the exercise in Table 4.43. It is particularly enlightening that in this exercise, almost 40 percent of the 13-year-olds and 25 percent of the 17-year-olds took two of the numbers and performed a single operation to arrive at their solution.

Table 4.43

Multistep Word Problem

Exercise	Percent Correct	
	Age 13	Age 17
Lemonade costs 95¢ for one 56-ounce bottle. At the school fair, Bob sold cups holding 8 ounces for 20¢ each. How much money did the school make on each bottle?*	11	29

*This exercise is similar to an unreleased exercise.

Summary

In closely examining the set of exercises involving decimals, it is apparent that many of the high levels of correct responses were made on exercises that required remembering a rule and not necessarily on understanding a decimal as a number. This was particularly true when coupled with those exercises with low levels of performance, such as those that required estimates, giving fraction-decimal equivalencies, ordering decimals, and dealing with thousandths. Although many students have facility with decimals, the foundation does not appear to be strong.

Percents

Overview of Results

Approximately 20 exercises were administered to 13- and 17-year-olds that dealt with basic concepts of percent with applications of percents. Overall performance on percent exercises was extremely low in these respects:

1. About one-third of the 13-year-olds and one-half of the 17-year-olds responded correctly to the basic concept exercises.
2. About one-sixth of the 13-year-olds and one-third of the 17-year-olds were successful on exercises that involved any sort of operation with percents or applications of percents.

3. Although students had difficulty with percent problems in general, both age groups did better when the percents were familiar ones, such as 25 percent.

Percent Concepts

Several exercises related the meaning of percent to "per hundred." The results of one exercise that asked students to express 9/100 as a percent are given in Table 4.44. The level of performance on this basic exercise is typical of all the exercises involving percent concepts. From the results on one unreleased exercise given only to 17-year-olds, it is evident that about two-thirds of them could interpret 35 percent as 35 out of 100.

Table 4.44

Percent Concept

Exercise	Percent Responding	
	Age 13	Age 17
Express $\frac{9}{100}$ as a percent.		
Acceptable responses		
9%, nine percent	36	53
Unacceptable responses		
.09, .09%	16	19
90, .9, .9%	11	9
11%, 11.1, .11, $\frac{1}{9}$	2	1
Other unacceptable responses	27	14
I don't know	8	4

Likewise, the same percentage of 13-year-olds realized that 100 percent makes a whole, in that they correctly responded to this exercise:

If 37 percent of the U.S. population is under 20 years of age, what percent of the population is 20 years of age or older?

When the whole was not partitioned into 100 parts, students had difficulty describing the figure in terms of percent. For example, in the exercise reported in Table 4.45, only 28 percent of the 13-year-olds and 53 percent of the 17-year-olds related the one-fourth of the shaded circles to 25 percent. Many more knew that one-fourth of the circles were shaded but failed to express the quantity in percent. For some students this error could have been caused by not reading or listening to the question, except for the one-fifth of the 13-year-olds and the few 17-year-olds that responded 2 percent (2 out of the 8 circles).

The ability to relate decimals, fractions, and percents was assessed by several exercises; the results of one are given in Table 4.46. Results on the other exercises were similar to these in that when the percent was a fairly common one, the level of performance was about the same as in parts A and C, but when the percent was not as common (e.g., .2 percent), then the level was more like that of part B.

Table 4.45

Expressing Quantity as a Percent

Exercise	Percent Responding	
	Age 13	Age 17

What percent of the circles is
shaded?

● ○ ○ ○

● ○ ○ ○

	Age 13	Age 17
Acceptable response, 25% or 25	28	53
Unacceptable responses		
2%	19	7
$\frac{1}{4}$, $\frac{2}{8}$, .25	38	29
.25% or $\frac{1}{4}\%$	1	2

Table 4.46

Relationship of Fractions, Decimals to Percents

Exercise	Percent Correct	
	Age 13	Age 17
A. Change 25% to a common fraction.	57	81
B. Change 125% to a decimal fraction.	27	44
C. Change .15 to a percent.	68	77

Percent Applications

One exercise that was not in a word problem context asked students to
perform operations with percents (see Table 4.47). The results show that
about one-third of 13-year-olds and over one-half of the 17-year-olds recognized that 30 is 50 percent of 60 (part A). The performance levels on the
remaining parts were extremely low. In another exercise students were able
to tell that 125 percent of a number would be more than the number, and they
also did reasonably well on exercises involving 25 percent. Yet they could
not put this knowledge together and decide what 125 percent of 40 would be.
If they needed to change 125 percent to a decimal to find a solution, then it
is easy to see why part D in Table 4.47 had such a low rate of success.

There were seven one-step application exercises, most of which required
finding a given percent of a number. The results of one such exercise given
in Table 4.48 are fairly typical. Although 10 percent of the 13-year-olds and
40 percent of the 17-year-olds responded correctly, almost one-third of the
13-year-olds tried to subtract or divide the two numbers. The one problem on
which students were more successful was a multiple-choice exercise that involved a common percent. About 46 and 69 percent of the 13- and 17-year-olds,
respectively, were able to select the correct answer. Any other variation in
the level of performance can be accounted for by the different difficulty
level of the multiplication involved.

Table 4.47

Operations with Percents

Exercise	Percent Correct	
	Age 13	Age 17
A. 30 is what percent of 60?	35	58
B. What is 4% of 75?	8	27
C. 12 is 15% of what number?	4	12
D. What is 125% of 40?	12	31
E. 6 is what percent of 120?	6	16

Table 4.48

One-Step Application of Finding a Percent of a Number

Exercise	Percent Responding	
	Age 13	Age 17
A store is offering a discount of 12 percent on winter coats. What is the amount a customer will save on a coat regularly priced at $30?*		
Acceptable response, $3.60, 3.60, 360¢	10	40
Unacceptable responses		
$26.40, (30 - $30 × .12), ($30 × .88)	2	3
30 - 12, 30 - .12	16	5
30 ÷ 12 = 2.5 or $2.50	15	9
$12, 12, .12, 12¢	4	2
Other	34	31
I don't know	11	6
No response	8	4

*This exercise is similar to an unreleased exercise.

One exercise asked what percent one number was of another (see Table 4.49). As was seen in previous examples, some students were able to give a correct decimal or common fraction relationship between the two numbers but did not express this relationship as a percent. Again, the somewhat higher level of response on this exercise may result from the familiarity of 25 percent.

The remaining exercises involved two or more steps, and, as could be expected, results were even lower on these than on the other application exercises. In one exercise (Table 4.50) students were asked to determine the percent of a discount. The most popular wrong answer, 12 percent, was the result of expressing the dollar amount saved, $12, directly as a percent. This process of interchanging dollars and percents was noted in several of the percent application exercises. The "add on" concept of percent, as represented in sales tax, has been generalized by many students to mean that a 5 percent increase or decrease in the cost of an item is the same as adding or subtracting $5 to or from the original cost. Fifteen percent of the students selected a 20 percent discount as the answer to the exercise in Table 4.50. It is likely that they added the $12 to the regular price, obtaining a total of $60, or perhaps they chose what they thought would be a reasonable discount.

Table 4.49

Finding What Percent One Number Is of Another

Exercise	Percent Responding Age 13	Age 17
A hockey team won five of the 20 games it played. What percent of the games did it win?		
Acceptable response, 25% or 25	21	52
Unacceptable responses		
.25, $\frac{1}{4}$, $\frac{5}{20}$	15	11
.4%, 4, .04	12	9
$\frac{1}{4}$%, .25%	1	2
75%, 75, .75, $\frac{3}{4}$	2	2
15%, 15	10	3

Table 4.50

Two-Step Percent Application

REGULAR PRICE $48.00

SALE PRICE $36.00

What is the percent of discount?

	Percent Responding Age 13	Age 17
○ 12%	52	34
○ 20%	14	15
● 25%	18	36
○ 33 $\frac{1}{3}$%	4	5
○ 75%	3	3
○ I don't know	9	7

Results on the other multistep exercises were similar to the results in Table 4.50. Although it is difficult to generalize, it appeared that over-all, exercises that involved finding a given percent of a number were easier than the other types of problems. As in most mathematics programs, finding a percent of a number was stressed more on this assessment. However, it does not appear from the results that students are successful on any of the basic types of percent problems.

Summary

The results of the set of exercises involving percents give reason for concern if one thinks that this is a topic that should be mastered at some level of facility by the age of 17. From the basic concepts through the ap-plications, there were weaknesses in the students' performance.

Other Number Topics

Overview of Results

In addition to the exercises involving whole numbers, fractions, decimals, and percents, there were sixteen exercises involving integers and ten exer-cises involving roots and exponents. Most of these exercises were adminis-tered to both 13- and 17-year-olds, but a few were given only to 17-year-olds. In general, the following statements about these exercises can be made:

1. The majority of the integer computation exercises involved small num-bers, and correct responses averaged about 40 percent for the 13-year-olds and 70 percent for the 17-year-olds. Performance on subtraction computation was much lower than these averages and was the lowest for any of the four operations.
2. The results on the exercises involving roots or exponents are best examined individually, since they vary greatly depending on the type of exercise.

Integers

Most of the integer exercises involved computation, but a few dealt with concepts and properties of these numbers. About 20 percent of the 13-year-olds and 70 percent of the 17-year-olds could select examples of positive num-bers that were expressed in terms of two negatives (e.g., $^-(-5)$). When asked to label an integer shown on a number line, 17-year-olds could label both positive and negative integers. Thirteen-year-olds were successful with the positive integers, but only 60 percent could label integers that were in the negative position on the line. Half of the 13-year-olds and over three-fourths of the 17-year-olds were able to order negative numbers. Very few at either age (14 and 16 percent) chose a verbal expression that could be represented by an integer. All the incorrect choices required a ratio or a fraction for de-scription, but the most common answer was "all of the above." Although this is only one exercise, it gives some indication that students have difficulty con-necting integers with real situations.

Table 4.51 summarizes the results on the integer computation exercises. As mentioned in the overview of results, most of the computation involved small numbers, such as those shown in the released subtraction and multiplica-tion exercises. Other than adding two positive numbers or two negative num-bers, there was little variation in performance within any one operation. It can be conjectured that either students remembered the rule for all cases or they did not know the rules for any of the cases. Further evidence for the validity of this conjecture is available through an exercise that asked stu-dents to decide when a product could be positive and when it could be negative. The same percentage correctly answered these questions as correctly found the product of two integers.

Table 4.51

Integer Computation

Exercise	Percent Correct	
	Age 13	Age 17
Addition Exercises*		
A. negative plus negative	75	80
B. negative plus positive	40	76
C. positive plus positive	90	93
D. positive plus negative	45	79
Word problem (negative plus positive)	54	78
Subtraction Exercises		
A. $^-4 - {}^+7$	21	39
B. $^-8 - {}^+10$	37	61
C. $37 - {}^-43$	23	47
D. $^-7 - {}^-10$	27	54
Word problem (difference between negative and positive)*	25	50
Multiplication Exercise		
A. $^-8 \times {}^+2$	49	78
B. $^+7 \times {}^-7$	36	65
C. $^-4 \times {}^-5$	34	73
D. $^+5 \times {}^-2$	39	67
Division Exercise*		
A. negative divided by negative	32	64
B. negative divided by positive	39	73
C. positive divided by negative (expressed as a fraction)	28	57
D. negative divided by negative (expressed as a fraction)	30	68

*Unreleased exercises

Note that on the whole, subtraction was more difficult than the other operations. The most common error was made by finding the difference between the two numbers while considering them as positive integers.

Results on the few word problems were consistent with the performance on similar computation exercises.

Roots and Exponents

The results were considerably higher for both 13- and 17-year-olds on exercises that required finding the square roots of perfect squares than they were when an approximation of a nonperfect square root was needed. For example, 60 and 80 percent of the 13- and 17-year-olds gave the correct response to such exercises as finding the square root of 36, but only about 10 and 60 percent chose the two integers between which the square root of 26 falls. A popular response (43 and 12 percent of the 13- and 17-year-olds) was that the square root of 26 was "between 26 and 27." Other ways of asking questions about approximations of square roots produced similar or lower results, suggesting that the skill of finding a square root is mechanical. The one exercise that asked for the fourth root (a whole number in this case) of a small number was correctly answered by 6 percent of the 13-year-olds and 43 percent of the 17-year-olds. One other exercise asked 17-year-olds to use a table to find the square root of 10n (15 percent were correct) and 100n (5 percent correct). In the latter use of the table, students had to realize that they were to find the square root of 100 times the number they had just found, but few realized this would multiply the original square root by 10.

Eighty percent of the 13-year-olds and 92 percent of the 17-year-olds interpreted 4^3 as $4 \times 4 \times 4$. Adding two numbers, each of which was expressed in terms of cubes, was slightly more difficult than multiplying two numbers, each expressed in terms of squares. About 36 and 57 percent, respectively, of the 13- and 17-year-olds did the addition exercise, and 55 and 69 percent were successful on the multiplication exercise. The difference could be accounted for by the error of adding the two numbers before cubing them. Results on both of these exercises were 2 to 11 percentage points lower than on the previous assessment. Only 17 percent of the 17-year-olds gave a number equivalent to a number raised to a negative exponent.

One exercise asked 17-year-olds to choose the scientific notation for a number such as 5,821,000. Thirty percent chose the correct notation, but over half of them selected 5.821×10^3.

Summary

The results of this assessment indicate that the beginning rules or skills associated with integer computation are fairly well developed for 17-year-olds except for the operation of subtraction. Both older groups can find the square roots of perfect squares but have difficulty with approximating square roots. Likewise, they seem to know the definition of a positive, integral power, but they cannot always apply this knowledge.

5

Variables and Relationships

The second assessment contained approximately eighty-five exercises intended to measure students' ability to work with mathematical variables and relationships. Most of these exercises dealt with algebraic concepts and manipulations, and the rest dealt with basic variable concepts that are included in elementary and middle school mathematics curricula. The exercises can be categorized as follows: (1) variables in equations and inequalities, (2) variables used to represent elements of a number system, (3) functions and formulas, and (4) coordinate systems.

Data collection in the second assessment permitted the study of 17-year-olds' performance based on mathematics course background. Of the students surveyed, 67 percent reported completing one year of algebra and 35 percent reported taking at least a half year of second-year algebra. Since students are assessed at specified ages and not necessarily immediately after they have completed a course in algebra, the assessment was for many 17-year-olds an inventory of algebraic skills and understanding retained one to two years after studying elementary algebra.

As would be expected, 17-year-olds' performance was clearly related to course background. Students who had taken one year of algebra scored about ten points above the mean for all 17-year-olds, and scores of students who had taken a second course in algebra were consistently fifteen to twenty points above the average for all 17-year-olds. Although it is reasonable to conclude that students who have taken additional algebra courses have learned more algebra as a result of this experience, it should also be kept in mind that generally the more capable students take a second course in algebra. Consequently, there is a selection process involved as well as additional instruction. It is also true that second-year algebra students are likely to have had more recent experience with algebraic skills than students who have not taken such a course.

A consistent cluster of 15 to 20 percent of the 17-year-old population showed mastery of algebraic skills and concepts as evidenced by the average performance levels on the most difficult exercises. A second group, about 30 to 40 percent of the population, showed an intuitive knowledge of algebraic processes. This group faltered when specific well-defined procedures were required.

There is little evidence to suggest any change in students' performance in algebra since the first assessment. The number of algebra exercises repeated on the second assessment is too small and the diversity too great to detect patterns in performance. For most repeated exercises, performance appeared stable for all 17-year-olds.

Equations and Inequalities

Overview of Results

Exercises assessing students' ability to solve equations included both simple open-sentence problems that are usually included in elementary and middle school mathematics curricula as well as more complex problems that required formal knowledge of algebra. The problems that required knowledge of formal algebraic operations included linear equations and inequalities in one unknown, systems of equations, and quadratic equations. Students were also asked to write equations to represent verbal problems and to solve typical textbook problems that required algebraic solutions. Basic concepts of inequality relations and formal algebraic solutions to inequalities were assessed. Following are the major results for which supporting data will be presented:

1. The 13-year-olds could solve simple linear equations intuitively, but they did not appear to know formal procedures for finding solutions to more complex equations.
2. Less than one-half of the 17-year-olds with a year of algebra could systematically solve linear equations; two-thirds of those with two years of algebra could do so.
3. About 25 percent of the 17-year-olds with a year of algebra and 40 percent of 17-year-olds with two years of algebra could solve simple systems of equations and quadratic equations by factoring.
4. About 10 percent of the 17-year-olds who had taken one year of algebra and 20 percent of those who had taken two years of algebra could solve a quadratic equation using the quadratic formula. Fifteen and 25 percent, respectively, made computational errors when they attempted to use the formula.
5. Performance on algebra word problems was consistently low.
6. Although most students were familiar with inequalities, most of them did not understand the special properties of inequalities and appeared to treat inequality relations as equalities.

Equations in One Unknown

Most elementary and middle school students have had some experience solving simple open sentences, and the results presented in chapter 4 describe students' performance in solving open sentences. The results suggest that almost all students could solve simple open addition sentences that could be solved by inspection, but subtraction sentences and sentences that required some sort of transformation because of the size of the numbers were generally more difficult.

The results summarized in Table 5.1 offer some additional insights into students' understanding of variables and their ability to solve simple equations. The results suggest that the format of the problem significantly affected performance. Students appeared to be more familiar with using a box to represent a variable than they were with using a letter.

Table 5.1

Simple Multiplication Equations

Exercise	Percent Correct Age 13
A. $4 \times \Box = 24$*	91
B. $6m = 36$*	65

*Similar to an unreleased exercise

The results for several exercises that could not be solved by inspection or a single transformation are summarized in Table 5.2. Since these exercises required some knowledge of algebraic operations, results have been summarized not only for all 17-year-olds but also for students who have taken one or two years of algebra. With the exception of the equation with nonnumerical coefficients (part D), there was relatively little difference in difficulty among the exercises even though they appear to differ in complexity and the number of steps required for solution.

Table 5.2

Linear Equations in One Unknown

		Percent Correct	
Exercise	Age 17	1 Year Algebra	2 Years Algebra
A. $9 = 5x + 2^*$	40	55	76
B. $30 = \frac{2}{5}c + 10$	33	43	62
C. $3x + 6 - 14 = x + 2$	34	44	63
D. Solve for x: $ax + b - 2 = d^*$	25	35	56

*Similar to an unreleased exercise

Systems of Equations

The levels of performance on an exercise that required a simple substitution to solve a system of equations are summarized in Table 5.3. Given the relatively straightforward substitution required, these results seem low for students with one or two years of algebra. Levels of performance are about 20 percentage points below students' performance on solving one equation with one unknown. Furthermore, performance on a somewhat more difficult exercise that could be solved by either substitution or addition was about 10 percentage points lower across all groups than the results for the exercise presented in Table 5.3.

Table 5.3

Linear Equations in Two Unknowns

Exercise			
$n = 3k$ What is the solution for this system of equations?			
$n + k = 72$ $k =$ _____ $n =$ _____			

		Percent Responding	
Response	Age 17	1 Year Algebra	2 Years Algebra
Acceptable response			
k = 18 and n = 54	19	26	40
Unacceptable responses			
k = 18, but n = 3 times value for k	3	4	5
k + n = 72, except as above	6	5	5
Only correct value for k or n	1	2	2
Other	36	41	33
No response or I don't know	30	22	14

Quadratics

The capstone of the traditional first course in algebra is solving quadratic equations. Student performance with quadratics was low. Although 60 percent of the 17-year-olds who had taken one year of algebra and three-fourths of those who had taken two years of algebra could identify a quadratic equation, they were less successful at solving even simple quadratic equations. The results summarized in Table 5.4 reveal some interesting insights about students' ability to solve quadratic equations by factoring. Exercises B and C involve subskills or concepts that are required to solve the quadratic equation in exercise A. There was relatively little difference in performance on the three exercises, although a larger percentage of those students with two years of algebra could solve the equation of exercise A than were successful on exercises B and C. The pattern of results suggests that students' difficulty with quadratic equations may stem from an inadequate knowledge of the required subskills.

Table 5.4

Quadratics

| Exercise | Percent Correct | | |
	Age 17	1 Year Algebra	2 Years Algebra
A. Find the solution set of $x^2 - 5x + 6 = 0$	18	26	53
B. What are the factors of $x^2 - 5x + 6$	22	28	43
C. Solve this equation for x $(x + 6)(x - 5) = 0$*	17	25	42

*Similar to an unreleased exercise

Students had even more difficulty using the quadratic formula to solve quadratic equations. Although almost half of the 17-year-olds with two years of algebra could identify the quadratic formula in a multiple-choice recognition exercise, few of them could apply it to solve a quadratic equation (Table 5.5).

Table 5.5

Using the Quadratic Formula

| Exercise | Percent Responding | | |
	Age 17	1 Year Algebra	2 Years Algebra
Use the quadratic formula $x = \dfrac{-b \pm \sqrt{b^2 - 4ac}}{2a}$ to solve the following equation: $2x^2 - 4x - 3 = 0$*			
Correct response, $\dfrac{2 \pm \sqrt{10}}{2}$	7	10	19
Correct substitution into formula, but made error or incomplete	6	9	17
Correct number substituted, but missed a sign or signs	3	5	8
Other	28	34	36
I don't know	40	31	13
No response	14	11	7

*Similar to an unreleased exercise

Translations into Equations

The use of algebra in most applications requires the translation of information into algebraic form. One of the most common experiences students face is that of translating the information provided in various situations into equations. The foundation for this skill is developed at the elementary school level by having students write open sentences to represent simple problem situations. Exercises assessing this skill are discussed in chapter 4. The results summarized in chapter 4 suggest that although students were reasonably successful at solving simple one-step verbal problems, they had more difficulty writing open sentences to represent problem situations. The results summarized in Table 5.6 offer further evidence that many students have not learned to represent simple arithmetic relationships using mathematical symbolism, especially when variables are involved.

Table 5.6

Algebraic Representation

Carol earned D dollars during the week. She spent C dollars for clothes and F dollars for food. Write an expression using D, C, and F that shows the number of dollars she had left.

| | | Percent Responding | | |
	Age 13	Age 17	1 Year Algebra	2 Years Algebra
Acceptable responses $D - (C + F)$ or $D - C - F$	17	45	57	67
Unacceptable responses $D - C + F$ or $D - F + C$	2	6	6	4
Other	51	33	33	26
No response or I don't know	30	16	4	3

Given that students had difficulty writing expressions to express simple arithmetic relationships, it is not surprising that solving algebra word problems was one of the most difficult algebra topics assessed. The results for two typical algebra word problems are summarized in Table 5.7. For exercise B, fewer than 4 percent of the students who had taken two years of algebra and only 2 percent of those who had completed one year of algebra wrote an appropriate equation to find their solution. It is impossible to determine whether students who correctly solved the problem did not write an equation because they were unable to do so or whether they felt it was not necessary to do so to find the answer. In any event, applying algebra skills to solve problems appears to be a major area of difficulty for most students. Although this is not greatly surprising, the little improvement in performance shown between the first course in algebra and the second was surprising. It appears that although additional study in algebra may improve students' algebraic skills, it does little to help them learn to apply those skills to solve problems.

Inequalities

The assessment included a number of exercises that focused on basic concepts of inequalities that might be expected of all students, whether or not they had studied algebra. These ideas included the use of appropriate notation, graphing inequalities on a number line, basic properties of inequality, and solving problems involving inequalities. The assessment also included exercises assessing students' ability to solve inequalities of the types covered

Table 5.7

Algebra Word Problems

Exercise	Age 17	Percent Correct 1 Year Algebra	2 Years Algebra
A. One number is 3 times as large as a second number. The sum of the two numbers is 72. What are the two numbers?	28	38	49
B. A supermarket charges $5.10 for a six-pound package of meat for a meatloaf. The package contains ground beef and ground pork. If ground beef sells for 80 cents a pound and ground pork sells for 95 cents a pound, how many pounds of ground beef are in the package?	20	25	31

in a course in algebra.

Most students had some familiarity with inequalities, but they lacked a clear understanding of the special characteristics of inequalities and appeared to treat inequalities as equations. Seventy-five percent of the 9-year-olds could use the <u>less than</u> symbol (<) correctly. The results summarized in Table 5.8 are representative of performance on several exercises assessing students' understanding of the transitivity of inequality.

Table 5.8

Properties of Inequality

Exercise	Percent Responding Age 9	Age 13	Age 17
If a > 5 and b > 5, then			
⊂ a equals b	31	32	16
⊂ a is greater than b	10	4	2
⊂ b is greater than a	4	3	2
● There is not enough information to determine the relation between a and b.	27	57	76
⊂ I don't know	16	5	3

In general, students encountered significantly more difficulty with inequalities than they did with equations of corresponding difficulty. Twenty-six percent of the 9-year-olds and 66 percent of the 13-year-olds could solve the following inequality:

Only 44 percent of the 13-year-olds and 69 percent of the 17-year-olds could identify the appropriate graph of the solution to a simple inequality on the number line.

Performance on several inequalities requiring algebraic solutions are summarized in Table 5.9. These results clearly illustrate how students attempted to treat inequalities just like equations. In exercise A, students were about as successful with the inequality as they were at solving the corresponding equation. In this problem the transformations involved are identical to those used to solve the equation, and the inequality remains unchanged. In exercise B, however, most solutions involved multiplying by a negative number, which would reverse the inequality. About 40 percent of the students failed to reverse the direction of the inequality.

Table 5.9

Solving Linear Inequalities

| | | Percent Responding | |
Exercise	Age 17	1 Year Algebra	2 Years Algebra
A. Which of the following inequalities is equivalent to $3x + 7 < 25$?*			
Correct response	58	70	82
B. Which of the following inequalities is equivalent to $5 - 2x < 13$?			
○ $x < -4$	13	17	22
○ $x < 4$	22	25	20
● $x > -4$	20	24	33
○ $x > 4$	12	15	14
○ I don't know	31	19	8

*Similar to an unreleased exercise

Students' superficial response to inequality problems and their failure to take into account the special characteristics of inequalities is further illustrated by the results in Table 5.10.

Table 5.10

Problem Solving Estimation

Ms. Baker has between $8000 and $8500 in her savings account. She wants to buy a new car that costs between $5300 and $5400. After she buys the car, how much money will Ms. Baker have in her savings account?

| | | Percent Responding | | |
	Age 13	Age 17	1 Year Algebra	2 Years Algebra
○ $2700	4	2	1	0
○ $3100	13	6	4	2
○ Between $2700 and $3100	59	64	66	64
● Between $2600 and $3200	21	27	28	34
○ I don't know	2	1	1	1

Summary

Most students at all ages assessed could solve simple open sentences that could be solved by inspection. The exercises that required formal algebraic skills clustered into three general levels of difficulty. About 40 percent of the 17-year-olds who had completed one year of algebra and 60 percent of those who had taken two years of algebra could solve linear equations in one unknown. The next most difficult cluster included systems of equations and quadratic equations that could be solved by factoring. About 25 percent of students with one year of algebra and 40 percent of students with two years of algebra could solve exercises of these types. The most difficult exercises involved solving quadratic equations using the quadratic formula and the application of algebraic skills to solve verbal problems. About 20 and 25 percent of the students with one and two years of algebra, respectively, were able to solve these exercises. Although most students had some familiarity with inequalities, they lacked a clear understanding of the special characteristics of inequalities and appeared to treat the inequality relation as if it were an equality.

Operations with Variables

Overview of Results

There were two major sets of exercises within this category: manipulation of algebraic expressions and the use of variables to represent basic mathematical concepts. The first group included simplification of linear or rational expressions, factoring, exponents, and radicals. The second focused on students' ability to use variables to express number theory concepts and to represent basic number properties and relationships. Following are the major results:

1. Students' success in simplifying algebraic expressions paralleled their understanding of the arithmetic concepts represented by the algebraic expressions.
2. Students generally could simplify expressions involving positive exponents but had difficulty with negative exponents and radicals.
3. Many younger students had difficulty using variables to express mathematical relations, but over 50 percent of the 17-year-olds could use variables to express most simple mathematical relations.

Simplifying Algebraic Expressions

On three unreleased exercises similar to those shown in Table 5.11, students' performance at all age levels decreased as the expressions became more complex. Over one-half of the 17-year-olds and nearly all students with two years of algebra chose the correct answer for the first two expressions. In the second exercise, the most common error resulted from the interpretation of $(1 + 5x)$ as $6x$. Except for students with two years of algebra, the performance level on the last exercise was about half that on the first exercise.

Table 5.11

Simplifying Linear Expressions

		Percent Correct	
Exercise	Age 17	1 Year Algebra	2 Years Algebra
A. $57 + 8x + 7y$*	64	82	91
B. $9(1 + 5x) + 3$*	52	67	85
C. $7 - 8(3 - y)$*	34	46	64

* Similar to an unreleased exercise

Students showed little ability to manipulate relational expressions. As shown in chapter 4, about 60 percent of all 17-year-olds could add two fractions with unlike denominators, but this ability did not transfer to expressions involving variables (Table 5.12). From the data shown, it is clear that three-fourths of the 17-year-olds resorted to adding the numerators and adding the denominators. The first foil listed represents the choice of nearly half of all algebra students. The second foil represents a second attempt at the same process, $3 + 3c = 6c$ and $2a + 4b = 6ab$. This response was more popular among students with little algebra. Apparently students did not recognize that the processes appropriate for adding fractions also apply to rational expressions involving variables.

Table 5.12

Rational Expression Addition

Exercise		Age 17	Percent Responding 1 Year Algebra	2 Years Algebra
$\dfrac{3}{2a} + \dfrac{3c}{4b} =$				
○	$\dfrac{3 + 3c}{2a + 4b}$	40	49	48
○	$\dfrac{6c}{6ab}$	34	24	13
○	$\dfrac{3 + 3c}{8ab}$	7	7	6
●	$\dfrac{6b + 3ac}{4ab}$	14	17	30
○	I don't know	5	3	2

Exponents and Radicals

Seventy percent of 17-year-olds could choose the exponent in the monomial expression $3x^5$. Two-thirds of the students with one year of algebra and nearly 90 percent of the students with two years of algebra could simplify an expression like $A^3 \cdot A^4$. However, student performance on a comparable division exercise illustrated a wide variety of interpretations. The results shown in Table 5.13 suggest that students had a strong tendency to perform a fraction simplification on the exponents alone, $4/20 = 1/5$. The fact that the correct choice was given as an expression with a negative exponent rather than as $1/a^{16}$ may have increased students' confusion, but even allowing for this possibility, the results show poor student understanding of powers and exponents when variables are involved.

Students also had difficulty with radicals (Table 5.14). Although three-fourths of the students with two years of algebra could simplify the expression in exercise A, they had a great deal more difficulty with exercise B. In fact, the more mathematics a student had taken, the more likely it was that the student would take the square root of the exponent. The results of this exercise show a serious confusion of powers and roots when radicals are involved.

Table 5.13

Use of Exponents

Exercise/Response	Age 17	Percent Responding 1 Year Algebra	2 Years Algebra
$\dfrac{a^4}{a^{20}} =$			
○ $\dfrac{1}{a^5}$	28	30	25
○ a^{-5}	14	16	14
● a^{-16}	22	28	43
○ a^{16}	7	6	6
○ a^{24}	6	3	2
○ $\dfrac{1}{5}$	6	6	4
○ $\dfrac{1}{5a}$	6	5	2
○ I don't know	9	4	1

Table 5.14

Simplification of Radicals

Exercise/Response	Age 17	Percent Responding 1 Year Algebra	2 Years Algebra
A. Which of the following is equal to $\sqrt{18}$?*			
○ $\sqrt{2}$	2	2	2
● $3\sqrt{2}$	44	58	78
○ $9\sqrt{2}$	31	23	14
○ 9	14	9	4
○ I don't know	8	7	2
B. What does the expression below equal?* $\sqrt{b^{36}}$			
Acceptable response			
b^{18}	11	16	26
Unacceptable responses			
b^6	27	35	47
6 or $b^2 = 36$	7	6	2
18 or 36	4	3	2
b^{36}	3	2	1
Other	12	23	15
I don't know	26	15	7

*Similar to an unreleased exercise

Summary

Students' ability to operate with variables was dependent on the complexity of the operations involved. In general, students had significantly more difficulty simplifying expressions that included variables or using variables to express mathematical relations than they did when only numerical operations were involved.

Using Variables to Express Mathematical Concepts

Variables provide a convenient notation to express general mathematical properties and relationships. Furthermore, understanding formulas and using algebra to solve problems depends on the ability to use variables to describe particular situations and express specific relations between terms. As would be expected, students' ability to represent particular relations using variables depended on their understanding of the relations involved and on the amount of algebra they had taken.

Several exercises required students to use variables to represent basic properties of numbers. All of these exercises are unreleased, but they were similar to the following problem:

> Which of the following expressions express the idea that when a number is multiplied by zero the result is zero?

Choices might include $x + 0 = 0$, $x \cdot 0 = 0$, and $x \cdot 0 = x$. Success on such problems ranged from 35 to 65 percent correct at age 13 and 50 to 70 percent at age 17.

While most students could identify equivalent fractions in which numerators and denominators were whole numbers, algebraic notation made that task more difficult. About one-third of the 17-year-olds (Table 5.15) recognized equivalent fractions involving variables, and about one-fourth of them responded "I don't know." Forty-five percent of students with one year of algebra and only slightly more than half with two years of algebra responded correctly.

Table 5.15

Equivalent Rational Expressions

Exercise	Age 17	Percent Responding 1 Year Algebra	2 Years Algebra
Which of the following is true for all numbers a and b, as long as b is not 0?*			
○ $\frac{a}{b} < \frac{5a}{5b}$	27	30	29
○ $\frac{a}{b} > \frac{5a}{5b}$	8	8	7
● $\frac{a}{b} = \frac{5a}{5b}$	37	45	54
○ I don't know	25	16	9

*Similar to an unreleased exercise

Functions and Formulas

A variety of different exercises assessed students' knowledge of functions both with and without formal notation. Formulas, which represent an important application of algebra, are applications of the function concept. A number of exercises were included that involved the evaluation and manipulation of formulas. The following represent major conclusions that can be drawn from the data:

1. Many students demonstrated some intuitive knowledge of functions, but they were less successful in using functional notation or using variables to describe functional relationships.
2. Students at all levels had difficulty manipulating formulas except when numeric data were substituted and the dependent variable was already isolated.

Functions

In learning mathematics, students may develop an intuitive concept long before they can handle an abstract representation of that concept. Many students demonstrated some intuitive notion of functional relationships but did not understand abstract representations that used variables or functional notation. Parallel exercises were administered to test evaluation of an expression like a + 7 with and without functional notation. The results are reported in Table 5.16. Students at all levels were able to evaluate the expression, but the exercise was more difficult when it was written as a function. One-third of those with no algebra and one-half of those with elementary algebra responded correctly.

Table 5.16

Evaluating Functions

			Percent Correct	
Exercise	Age 13	Age 17	1 Year Algebra	2 Years Algebra
A. What is the value of a + 7 when a = 5?*	70	86	96	98
B. If f(a) = a + 7, what is the value of f(5)?*	29	43	53	65

*Similar to an unreleased exercise

A similar exercise is reported in Table 5.17. In this exercise, students were required to find the rule or function from the table by first identifying a specific entry of the table and then expressing the rule using a variable. Some students had difficulty generalizing the rule from the table, but many more were unable to express the generalization using a variable. These results provide another example of how students are unable to use variables to express relations that they appear to understand when only numerical operations are required.

Table 5.17

Identifying a Function

x	y
1	8
3	
4	11
7	14
n	

Exercise		Percent Correct		
	Age 13	Age 17	1 Year Algebra	2 Years Algebra
A. What is the value of y in the above table when x = 3?*	35	59	69	81
B. What is the value of y in the above table when x = n?*	3	31	41	58

*Similar to an unreleased exercise

Formulas

An important application of algebra is the expression of relationships in other subject areas by mathematical formulas. Several exercises assessed students' ability to interpret algebraic formulas. Other exercises assessed students' skill at manipulating formulas.

On several exercises, students were asked to identify relationships presented in a formula. The exercise shown in Table 5.18 was administered to both 13- and 17-year-olds. Somewhat surprisingly, performance for both age groups was comparable, and the results for algebra students were only slightly higher than results for all 17-year-olds.

Table 5.18

Formula Relationships

Some people suggest the following formula be used to determine the average weight for boys between the ages of 1 and 7:

$$W = 17 + 5A$$

where W is the average weight in pounds and A is the boy's age in years. According to this formula, for each year older a boy gets, how much more should he weigh?

	Age 13	Percent Responding Age 17	1 Year Algebra	2 Years Algebra
● 5 pounds	54	54	58	64
○ 17 pounds	21	22	22	19
○ 22 pounds	17	14	13	11
○ I don't know	8	9	7	6

However, students' ability to substitute values into formulas appeared to be related to their algebra background (Table 5.19). For all students, though, the location of the dependent variable in the formula was critical to success. For a formula like W = 17 + 5A, about half to two-thirds of the students had success in finding W when given the value for A. Performance was much lower when a value for A was to be found. In the example in Table 5.19, only about one in three who responded correctly in part A could solve for a variable embedded in the formula.

Table 5.19

Formula Substitution

The formula for the efficiency of an engine is:

$$e = \frac{m}{g}$$

where e is the efficiency, m is the mileage, and g is the gallons of gasoline.[*]

| | | Percent Correct | | |
	Age 13	Age 17	1 Year Algebra	2 Years Algebra
A. The mileage of a trip was 112 miles and the engine used 15 gallons. What is its efficiency?	33	66	79	91
B. The efficiency of an engine is 12 miles per gallon. How many gallons of gas are needed for a trip of 9 miles?	6	19	27	38

*Similar to an unreleased exercise

The poor results in formula manipulation were consistent on several exercises including a common physical science formula and a metric conversion relationship. When asked to solve for a variable not isolated in the formula, fewer than one-fourth of the nonalgebra students responded correctly; one-half or fewer of those with two years of algebra were correct.

Coordinate Systems

Overview of Results

Seventeen-year-olds were asked to identify and locate points on a coordinate system, draw lines on a coordinate system, identify the slope and intercepts of lines, and find the equation of a line from its graph. Several exercises were included that attempted to assess younger students' understanding of certain concepts underlying the development of formal coordinate geometry. Following are the major results:

1. There was a steady growth in students' ability to locate and label points in a two-dimensional coordinate plane.
2. Seventeen-year-old students have failed to master the relationships between linear equations and their graphs.

Results

One common application of coordinate systems is to locate towns on a map. In general, even younger students were successful at this task. On a map with one axis lettered and the other axis numbered, 80 percent of 9-year-olds and well over 90 percent of older students chose the town located near a given grid position.

The exercise presented in Table 5.20 attempted to assess how well younger students could deal with the skills involved in graphing points on a coordinate system. The exercise was designed to have enough information so that even students who had not had instruction on coordinate systems could answer the question. Many 9-year-olds tended to disregard the order implied by the ordered pair. By age 13, most students were successful with this task.

Table 5.20

Labeling and Plotting Points

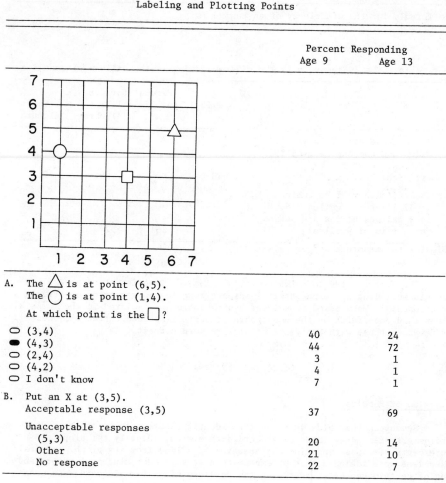

	Percent Responding	
	Age 9	Age 13
A. The △ is at point (6,5). The ◯ is at point (1,4). At which point is the ▢?		
(3,4)	40	24
● (4,3)	44	72
(2,4)	3	1
(4,2)	4	1
I don't know	7	1
B. Put an X at (3,5). Acceptable response (3,5)	37	69
Unacceptable responses		
(5,3)	20	14
Other	21	10
No response	22	7

Thirteen-year-olds were less successful at locating points on a standard coordinate system, but by the age of 17 most students had mastered this skill. About 40 percent of the 13-year-olds and 75 percent of the 17-year-olds could identify a point in the third quadrant.

For many applications, the relationship between linear equations and their graphs is of fundamental importance. While students could graph ordered pairs on the xy-plane, they did not show a knowledge of the relationship between equations and their graphs. Only 18 percent of the 17-year-olds correctly graphed a linear equation given a ruler and sheet of graph paper with labeled axes. One-fourth of the students with one year of algebra and fewer than half of those with two years of algebra were successful. Given a graph of a straight line with clearly marked intercepts like (-3, 0) and (0, 5), 5 percent of the 17-year-olds and about one-fifth of the students with two years of algebra could write the equation.

Given $2y = 5x - 8$, about 15 percent of the 17-year-olds could determine the slope of the line represented by the equation. Twenty percent of the students with one year of algebra found the correct slope; 35 percent of the students with two years of algebra were successful. Performance was slightly lower when students were asked to find the y-intercept. As in solving equations, the major differences in the response patterns among students of different algebra backgrounds (other than the correct answer) were the "I don't know" and no response options.

Summary

Despite some graphing activities in algebra texts, students' knowledge of the relationship between linear equations and graphing does not seem well established. To help develop this relationship, it might be necessary to do more with the graphing of linear data from science or social studies. The study of meaningful data in tables, graphs, and equations simultaneously might help students internalize the relationships.

Trigonometry

Only three exercises dealing with trigonometry were included in the assessment. Performance was uniformly low, suggesting that most students have little knowledge of the most fundamental trigonometric concepts. Given a right triangle with its sides labeled, only about 10 percent of the 17-year-olds could express the sine, cosine, or tangent of a given angle. Only 7 percent could find the length of a side of a right triangle using basic trigonometry.

6

Geometry

The seventy-eight geometry exercises administered during the second mathematics assessment dealt primarily with geometric ideas that students would have encountered outside a formal course in deductive geometry. The exercises can be classified into four major categories: (1) shapes, (2) shape and size relationships, (3) geometric constructions, and (4) problem solving. Most of the exercises were contained in category 1. Twenty-three exercises were released; eight exercises were repeated from the first assessment.

Shapes

Overview of Results

The exercises in this group assessed students' ability to recognize plane and solid shapes and their knowledge of the properties of those shapes. Following are the major results:

1. Most students in all three age groups were able to recognize simple plane figures, and the older respondents were able to recognize simple three-dimensional shapes that used common vocabulary.
2. For the most part, students were less successful on exercises that assessed knowledge of the properties of geometric figures than they were on recognition exercises.
3. Seventeen-year-olds who had studied a year of geometry (46 percent of the sample) generally scored much higher than 17-year-olds with no geometry course.

Recognition of Figures

Table 6.1 summarizes the results of several recognition exercises. As Table 6.1 shows, almost all students assessed were able to recognize simple plane figures that they encounter as part of everyday experience. Other recognition tasks assessed students' familiarity with common solid objects, including both their ability to select the name of a solid when given an everyday representation of the object and their ability to select the diagram that represented a named solid.

The naming exercise had been administered during the first assessment, and as Table 6.2 shows, performance levels between assessments showed minor changes. Although only about one-fourth of the 9-year-olds could correctly name a cylinder and a sphere, approximately two-thirds of the 13-year-olds and over 80 percent of the 17-year-olds could do so.

The results shown in Tables 6.1 and 6.2 suggest that the combination of school instruction and real-life situations apparently provides the kinds of experiences students need in order to become familiar with basic geometric shapes. As expected, there was a noticeable difference between recognition and use of technical terms, such as sphere and perpendicular, and more common everyday terms, such as cube and parallel.

Table 6.1

Geometric Recognition Tasks

Figure	Percent Correct		
	Age 9	Age 13	Age 17
Triangle	88*	85	86
Square	93	96	96
Rectangle	84	92	--
Circle	99	--	--
Geometric plane	21	50	73
Parallel lines	57	90	94
Perpendicular lines	15	33	70
Radius of circle	--	57	78
Chord of circle	--	33	55
Tangent of circle	--	35	59

*The 9-year-olds had a different exercise from that given to the older respondents.

Table 6.2

Identifying Common Solids

Correct Response	Percent Correctly Selecting								
	Name of Object						Diagram of Object		
	Age 9		Age 13		Age 17		Age 9	Age 13	Age 17
	1st	2nd	1st	2nd	1st	2nd			
Rectangular solid	49	48	82	79	94	92	--	--	--
Cylinder	21	24	66	66	85	85	41	78	93
Sphere	29	29	69	67	87	82	20	65	79
Cube	--		--		--		85	96	99

Results on other recognition exercises indicate that knowledge of technical terms is also related to the taking of a geometry course. For example, 57 percent of the total 17-year-old sample correctly recognized the hypotenuse of a right triangle; however, 91 percent of the 17-year-olds who had taken a year of geometry knew the term, as compared to only 20 percent of the students who had not taken geometry.

Properties of Figures

Triangles. Although students were able to recognize triangles, they demonstrated much less familiarity with some of the properties of triangles. Two unreleased exercises assessed respondents' knowledge that each side of a triangle must be shorter than the combined length of the other two sides. In one exercise, students were asked to select triples of numbers that could serve as lengths of the sides of a triangle. Over half of the 9-year-olds and almost two-thirds of the 13-year-olds were able to select triples of segments that could represent the sides of a triangle, but only 11 percent of the 13-year-olds and 18 percent of the 17-year-olds successfully selected triples of numbers that could serve as lengths of the sides of a triangle. Apparently, some students have some intuitive notion of this property that may be based on mentally constructing the triangle. They are unable, however, to extend this intuitive notion to an abstract situation, which requires explicit knowledge of the fact that the sum of the measures of two sides of a triangle must be greater than the measure of the third side.

Two unreleased exercises assessed students' knowledge of the fact that the sum of the measures of the interior angles of a triangle is 180 degrees. Between 10 and 20 percent of the 13-year-olds were successful on these exercises. Thirty percent responded "I don't know" to one of them, and 30 percent thought not enough information had been given to answer the question posed by the other.

The 17-year-olds were somewhat more successful on these exercises, with levels of performance ranging from 40 to 60 percent correct. Performance was related to whether or not the students had studied geometry. For those 17-year-olds who had studied geometry, success rates ranged from 68 to 86 percent correct; 9 to 32 percent of those 17-year-olds who had not studied geometry were correct.

Quadrilaterals and other polygons. One exercise dealt with 17-year-olds' knowledge of relationships among certain types of quadrilaterals, as shown in Table 6.3. The results of this exercise suggest that a majority of students knew that squares and rectangles are types of parallelograms (parts B and D) and, consequently, that it is not true that every parallelogram is a square (part A). They did not, however, recognize the relationship between squares and rectangles (part C).

Table 6.3

Relationships among Quadrilaterals

	Statement	Percent Responding (Age 17)		
		True	False	I Don't Know
A.	Every parallelogram is a square.	16	73*	11
B.	Every square is a parallelogram.	62*	27	11
C.	Every square is a rectangle.	32*	65	2
D.	Every rectangle is a parallelogram.	64*	25	10

*Indicates correct response

Another exercise presented to 13- and 17-year-olds asked for necessary and sufficient conditions for a quadrilateral to be a rectangle. The exercise and results are presented in Table 6.4. The success rate was low even for those 17-year-olds who had studied a year of geometry. It seems likely that students may not have had experience in recognizing or defining the properties of a rectangle.

Results for an exercise dealing with interior angles of a polygon are shown in Table 6.5. A large percentage of both age groups concluded that more information would be required to find the solution. Selection of this response may have been used as a substitute for an outright response of "I don't know," although there is no way to determine the accuracy of that conjecture. Furthermore, students may not have been sure of the meaning of the phrase interior angle given in the exercise.

Table 6.4
Determining Properties of a Rectangle

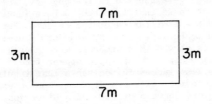

In the figure above, the lengths of the sides are shown. Which of the following guarantees the figure is a rectangle?

Response	Age 13	Age 17	Percent Responding Age 17 with Geometry	Age 17 without Geometry
⭘ The opposite sides are congruent.	26	31	40	23
● The opposite angles are congruent.	9	8	7	9
⭘ The angles are right angles.	14	20	29	12
⭘ The opposite sides are parallel.	40	34	22	44
⭘ I don't know and no response	11	7	1	11

Table 6.5
Finding Sum of Interior Angles of a Polygon

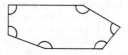

What is the sum of the interior angles of this plane figure?

Response	Age 13	Age 17	Percent Responding Age 17 with Geometry	Age 17 without Geometry
⭘ 540°	9	11	13	9
● 720°	7	13	18	7
⭘ 900°	2	3	2	3
⭘ 1080°	1	4	6	2
⭘ Not enough information given	67	56	58	56
⭘ I don't know and no response	15	13	4	23

75

Visualization Tasks

A group of exercises assessed students' ability to look at a particular diagram of an object and to visualize what would happen if some transformation (e.g., folding or cutting) were applied to the object. There was a wide range of success over different tasks that appeared to be a function of the complexity of the task and the familiarity of the object. Performance on many exercises also appeared to be influenced by students' out-of-school experiences.

In general, results showed that students were successful on tasks that required them to visualize how a shape would appear from a given perspective. They were progressively less successful on tasks that required them to abstract increasing amounts of information from a diagram. As an example of the first type of task, students were shown a picture of a block that had one corner removed. They were asked to select the diagram that would illustrate how the block would look when viewed from above. Approximately three-fourths of all three age groups responded correctly to this exercise. The most striking feature of the results was the remarkable consistency across age levels on each response choice.

Two unreleased exercises that generated contrasting results required the 9-year-olds to visualize what would happen if a certain shape were sectioned. The first of these asked them to visualize the figure that would be formed if a certain familiar object were cut open and flattened out. Thirty-six percent of the 9-year-olds gave the correct response. On the other exercises, however, 91 percent gave the correct response. In this exercise, students were shown a picture of a roll of clay and were told that it was to be cut lengthwise. They were asked to select the diagram that illustrated what one of the resulting pieces would look like. There may be several explanations for the differences in performance on what appear to be related tasks. The first exercise involved technical mathematical terms, and the response options were verbal; the second used no technical terms, and the response options were pictures. The higher performance on the second task may also result from students' experience in working with real objects. The task called for in the second exercise is one that students may actually have done, whereas it is likely that the first task was less familiar to the students.

A sectioning exercise given to 13- and 17-year-olds asked them to select the diagram of the solid from among four candidates that could not have been cut once to form a given shape. Forty-eight and 61 percent of the 13- and 17-year-olds, respectively, responded correctly. These levels of performance reflect the trend cited earlier of decreasing performance associated with increasing levels of abstraction. Not only was the exercise more complex than the pictorial sectioning exercise given to the 9-year-olds but the format of the question (which could not be cut to form the given shape) may also have contributed to the difficulty level of the exercise.

The final visualization exercise to be discussed asked students to construct a solid from its component parts. The 9- and 13-year-olds were shown four "stamps" that had resulted from a particular solid being dipped in paint, and were asked to select the object that had made the stamps from a group of four choices. Only 24 percent of the 9-year-olds and 43 percent of the 13-year-olds were successful; 66 percent of the 9-year-olds and 45 percent of the 13-year-olds selected a choice that could have made three of the four stamps. The task was not one commonly encountered in the elementary school mathematics curriculum, and the results indicated that this type of visualization task was the most difficult among those assessed.

Summary

Results of the exercises presented in this section have shown that most students of all three age levels were able to recognize simple plane and solid shapes but that they were less familiar with basic properties of those shapes. It was suggested that students' out-of-school experiences probably have a much larger influence on the recognition tasks than they do on students' knowledge of specific properties of geometric figures. Performance on some visualiza-

76

tion tasks also appeared to be influenced more by real-life experience than by
school instruction.

Shape and Size Relationships

Overview of Results

Exercises in this category dealt with students' knowledge of congruent
and similar figures along with their knowledge of some properties of congruence and similarity. Other relationships between figures were also assessed.
Following are the major results:

1. A large majority of students at all age levels were able to recognize congruent and similar figures, although language factors like the
   use of the term congruent influenced performance.
2. Although students were able to recognize congruent and similar
   figures, they were less knowledgeable about the properties of congruence and similarity.
3. Students were relatively unsuccessful on exercises that dealt with
   a knowledge of properties of other figure relationships, such as
   parallel lines, but having taken a geometry course affected performance levels on such exercises.

Congruence

Three exercises assessed students' ability to determine whether two given
figures were congruent. Eighty-two percent of the 9-year-olds and 94 percent of
the 13-year-olds correctly selected the pair of figures that "have the same
shape and size." Success rates were slightly lower on another congruence exercise that required a visual transformation of one of the figures. On this exercise, 54, 73, and 83 percent of the 9-, 13-, and 17-year-olds, respectively,
answered correctly.

The third exercise used the term congruent. The exercise and results are
represented in Table 6.6. Students were generally successful in identifying
the one figure that was congruent to the given rectangle. Results on parts C
and E suggest, however, that many students may lack a clear idea of congruence
and that they may base their notions of congruence on numerical relationships
without regard for size and shape.

Similarity

As with congruence, respondents were successful at the task of recognizing similar figures from a diagram but were much less successful in dealing
with the properties of similar figures. A typical recognition exercise is
presented in Table 6.7. This exercise was simple for the respondents, although undoubtedly students' everyday use of the term similar contributed to
the high level of performance. Furthermore, none of the pairs of figures were
reflections or rotations of each other. Performance might have been lower if
this had been the case. These results were comparable to another multiple-part
recognition exercise given to 13- and 17-year-olds in which from 77 to 90 percent of the 13-year-olds and 80 to 96 percent of the 17-year-olds responded
correctly to the different parts of the exercise.

Although the students were able to recognize parts of similar figures,
they demonstrated much less familiarity with the properties of similar figures.
Such an exercise is presented in Table 6.8. Students' lack of familiarity
with properties of similar figures helps to explain the relatively low levels
of performance on certain problem-solving exercises that will be discussed in
a later section.

Table 6.6

Basic Congruence Exercise

Is each figure below congruent to the figure above?

		Percent Correct		
	Age 13	Age 17	Age 17 with Geometry	Age 17 without Geometry
A.	65	71	83	58
B.	58	76	88	64
C.	33	28	37	19
D.	89	93	98	88
E.	50	50	60	40

78

Table 6.7

Recognition of Similar Triangles

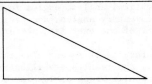

Which one of the following figures is SIMILAR to the above triangle? Fill in the oval beside the one you choose.

| | Percent Responding | |
	Age 9	Age 13
	9	3
	86	94
	1	1
	3	2
I don't know	1	0

Table 6.8

Properties of Similar Triangles

Two triangles are SIMILAR. Indicate if each one of the following statements is true or false.

Statement	Age 13	Age 17	Age 17 with Geometry	Age 17 without Geometry
A. Their corresponding sides MUST be congruent.	32	44	57	31
B. Their corresponding angles MUST be congruent.	57	68	83	53
C. They MUST have the same area.	45	65	80	52
D. They MUST have the same shape.	62	74	82	67

Other Relationships

Exercises in this group assessed students' knowledge of several topics, including adjacent angles, vertical angles, angles formed by a transversal cutting parallel lines, supplementary angles, and basic concepts of symmetry. With the exception of one part of one exercise, performance levels averaged less than 50 percent correct.

Table 6.9 presents the results of an exercise that dealt with students' knowledge of adjacent angles. The results are illustrative of other exercises in this category. The most popular wrong answer for both age groups was 90°. This response was probably based on the fact that rays BA and BX appear to be perpendicular, so that 90° was, in the students' view, a reasonable choice. These results illustrate a problem that teachers have long recognized—students' tendencies to make assuptions about a diagram. Two-thirds of the 17-year-olds who had taken a year of geometry responded correctly to this question, and so perhaps experience in a geometry course alerts students to the fact that they cannot depend on the appearance of a given figure.

Table 6.9

Results on an Exercise about Adjacent Angles

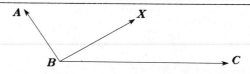

∠ ABC is obtuse. ∠XBC is 40°. What is the measure of ∠ABX?

Response	Percent Responding Age 13	Age 17
○ 40°	6	2
○ 50°	12	7
○ 90°	35	30
○ 140°	13	10
● Not enough information given	24	41
○ I don't know	10	10

Most of the 9- and 13-year-old respondents were successful on exercises that involved basic ideas of symmetry. For example, 78 percent of the 9-year-olds and 88 percent of the 13-year-olds could identify a figure that was not symmetric with respect to a given line, and 60 and 80 percent of the 9- and 13-year-olds correctly identified an unfolded version of a symmetric figure. However, only 21 percent of the 13-year-olds could complete a drawing with respect to two given lines of symmetry, a level of performance that most likely reflects a lack of experience with that type of task.

The effect of the choice of language on performance was dramatically illustrated by the results of two exercises given to the 9- and 13-year-olds. When asked to select a letter that could be folded so that both sides would match, 80 percent of the 9-year-olds and 83 percent of the 13-year-olds were successful; however, when the same questions were asked using the term symmetric, performance levels dropped to 18 percent and 20 percent correct for the 9- and 13-year-olds, respectively, with around 50 percent of both groups responding "I don't know."

Summary

The results presented in this section have shown again that students are successful in the lower-level recognition skills but are less successful in dealing with the properties of relationships. Out-of-school experience appeared to have little, if any, influence on students' knowledge of certain size and shape relationships, with the possible exception of the term <u>similar</u>.

Geometric Constructions

Overview of Results

Following is the major observation based on the results of the constructions exercises:

Seventeen-year-olds generally have difficulty with geometric constructions, but they are much more adept at recognizing proper constructions than they are at making those constructions.

Discussion of Results

Four exercises were given that assessed students' ability to recognize appropriate geometric constructions and to perform such constructions. Students were much more adept at recognizing constructions than they were at making those constructions on their own. For example, 53 percent of the 17-year-olds recognized the correct illustration of the bisection of an angle, but only 25 percent of them could actually bisect an angle using a compass and straightedge. For those 17-year-olds who had studied geometry for a year, the corresponding percentages were 70 and 48 percent correct; for those 17-year-olds with no geometry, results showed only 32 and 4 percent correct. Studying geometry obviously makes a difference but does not guarantee success on the task. Similar results were found on a pair of exercises with the task of constructing a perpendicular to a given line at a given point.

Problem Solving in Geometry

Overview of Results

In all content areas of the mathematics curriculum, the assessment results have shown that students consistently performed poorly on problem-solving tasks. Geometry is no exception to this generalization. Following are the major observations for which supporting data will be presented:

1. Fewer than two-thirds of the older groups were successful on any question that required them to make simple inferences.
2. The majority of 13- and 17-year-olds were not able to apply the Pythagorean theorem to solve routine problems, nor were they able to apply properties of similar triangles in problem situations.
3. Performance levels on geometric problem-solving tasks, although low, remained relatively unchanged from the previous mathematics assessment.

Inference Questions

Most of the questions on this portion of the assessment were administered only to 13- and 17-year-olds. The results of one exercise are shown in Table 6.10. Although performance on open-ended exercises is generally lower than it is on multiple-choice exercises, the relatively low levels of performance on this exercise are surprising. The way in which the question was worded—"separated into halves by a line parallel to the base"—might have confused some of the respondents. Also not clear is what caused the decline of 8 percentage points in performance for the 17-year-olds from the first to the second assessment. This is the only exercise in this category for which a decline this large was found.

Table 6.10

Results on an Inference Type of Question

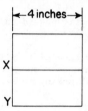

The square region above is separated into halves by a line parallel to the base side. What is the length of line segment XY?

| | Percent Correct | |
	Age 13	Age 17
First assessment	33	52
Second assessment	33	44

Other questions required 13- and 17-year-olds to apply their knowledge of vertical, supplementary, and adjacent angles in simple inference-type situations. Performance varied with the nature of the task, but it was generally low.

Pythagorean Theorem

Four exercises required application of the Pythagorean theorem for their solution. The tasks were, for the most part, routine textbook problems that required students to find the length of the hypotenuse of a right triangle. Three of the four triangles involved were either 3-4-5 triangles or a simple variant, such as a 9-12-15 triangle. In spite of the fact that a 3-4-5 right triangle provides probably the most familiar situation within which students are asked to apply the Pythagorean theorem, performance levels on these exercises were low, ranging from 18 to 49 percent correct for the 17-year-olds. Nineteen percent of the 13-year-olds were correct on the only such exercise given to them.

One Pythagorean theorem exercise given to the 17-year-olds also had been given during the first assessment. Seventeen percent of the 17-year-olds were correct on the first assessment of the exercise; 18 percent responded correctly on the second assessment. A somewhat similar exercise required the same task, but the exercise was illustrated by a diagram that might have facilitated the solution. Nineteen percent of the 13-year-olds and 37 percent of the 17-year-olds responded correctly. Only 54 percent of those 17-year-olds with a year of geometry solved the problem.

The only released exercise dealing with the Pythagorean theorem is shown in Table 6.11. In order to solve this problem, students had to know that a tangent to a circle was perpendicular to the radius at the point of tangency. Whether the students who solved the exercise correctly knew this or simply decided this must be a 3-4-5 right triangle is not clear. The 20 percent who responded "7 cm" might have either equated the distance from A to B with the distance from A to C or added 3 and 4 to obtain 7. The strategy of adding all

82

the numbers shown in an exercise to obtain a solution has been found repeatedly in this assessment. Again, taking geometry had some influence on the responses, with 63 percent of those students taking geometry responding correctly.

Table 6.11

Results on an Exercise Using the Pythagorean Theorem

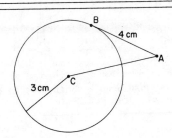

The tangent segment $\overline{AB}$ is 4 cm long. If the radius of the circle is 3 cm, how far is point A from the center of the circle?

Response	Percent Responding Age 17
⊂ 3 cm	3
⊂ 4 cm	5
● 5 cm	49
⊂ 6 cm	13
⊂ 7 cm	20
⊂ I don't know	11

Similarity

Given students' unfamiliarity with some of the basic properties of similarity, students' low performance on problems that required the application of those properties is not surprising. Responses to some questions showed a lack of knowledge of similarity properties.

Performance was somewhat higher on the typical textbook problem shown in Table 6.12. About two-thirds of those 17-year-olds with a year of geometry responded correctly. Somewhat atypical in the results is the fact that over twice as many 17-year-olds as 13-year-olds responded "I don't know" to the question.

Summary

Results on the exercises presented in this section indicate that students, for the most part, are unable to apply their knowledge to solve even routine problems that involve geometric ideas. The scope of the geometry curriculum outside of formal geometry courses has typically been limited to teaching students to recognize simple geometric figures and to acquaint them with the most basic properties of those figures. Results obtained on the geometry exercises given during the second mathematics assessment suggest that students do have knowledge of basic geometric concepts, and therefore, from this point of view, instruction in school geometry is apparently successful in meeting curricular objectives.

The emphasis on such lower-level skills, however, does not leave much time for instruction in geometric ideas that are potentially more valuable to students than simple recall and recognition tasks. For example, exploring relationships among geometric figures, such as quadrilaterals, not only teaches

Table 6.12

Results on an Exercise Involving Similarity

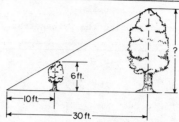

The picture above shows how Jose used a short tree to find the height of the tall tree. What answer should Jose get?

| Response | Percent Responding | |
	Age 13	Age 17
○ 12 feet	15	7
● 18 feet	36	50
○ 20 feet	9	9
○ 35 feet	15	11
○ 46 feet	17	9
○ I don't know	6	13

the content-specific information but also provides experience in the more general skills of classifying and organizing information that are useful in several areas of the curriculum. Emphasis on skills useful in solving geometric problems would also pay off in many areas. The geometry curriculum should be expanded to provide opportunities for students to learn such skills.

7

Measurement

Seventy-eight exercises tested measurement concepts. Twenty-two of these exercises were administered in the previous assessment and consequently provide some measure of change. Unfortunately, the change exercises are not evenly distributed over the complete range of measurement topics, and it is difficult to get a picture of change in measurement abilities as a whole.

With a few exceptions, the exercises fall into four basic content areas: (1) recognizing and comparing units, (2) measuring and estimating length, (3) time and reading measuring instruments, and (4) perimeter, area, and volume.

In general, it appears that students are successful with measurement skills that are reinforced outside of school. They can tell time and make simple linear measurements. They are familiar with common units of measure. However, they have difficulty with measurement concepts and skills that they do not encounter outside school. Furthermore, most of their measurement skills have been learned at such a superficial level that they cannot apply them to situations that are even slightly unfamiliar.

Recognizing and Comparing Units

Overview of Results

Three principal abilities involving units of measure were assessed: (1) selecting appropriate units, (2) recognizing a given unit and making simple estimates using a given unit, and (3) converting to different units of measure. All three abilities were assessed for both English and metric units.

The following general observations can be made regarding students' knowledge of standard units of measure:

1. Most students know the relationships among standard English units of measure but have difficulty applying this knowledge to solve conversion problems.
2. Knowledge of the relationships among standard English units has not changed since the first assessment, but there has been a significant decline on application problems. The poor performance on application exercises appears to reflect students' difficulty with problem solving in general.
3. Although performance was slightly lower on metric measurement exercises than on corresponding English measurement exercises, there has been a significant increase in students' familiarity with basic metric units since the time of the first mathematics assessment. About two-thirds of the 9-year-olds and most 13- and 17-year-olds have had at least some exposure to metric measurement. However, most students have not begun to "think metric," in that they are still not familiar with the magnitude of metric units and cannot make simple estimates using metric units.

English Units

At all ages, most students appear to be familiar with common units of measure. In the one exercise assessing students' ability to select appropriate

English units, 88 percent of the 9-year-olds identified _miles_ as the appropriate unit for measuring long distances. Students' familiarity with common units is also reflected in their ability to make rough estimates of length. Fifty-four percent of the 9-year-olds, 79 percent of the 13-year-olds, and 86 percent of the 17-year-olds correctly estimated the height in feet of a common object.

Students' ability to convert from one unit to another varied greatly depending on the units involved and the context of the problem. Students' knowledge of basic unit conversion factors is summarized in Table 7.1. The difficulty of the conversion between quarts and pints appears to indicate that students confuse the conversion of pints to quarts and quarts to gallons. Thirty-two percent of the 13-year-olds and 26 percent of the 17-year-olds chose 4 as the number of pints in a quart. Students' success with the foot-to-inches conversion may reflect more extensive actual experience in measuring with these units both in and out of school. The fact that measurements involving feet and inches are frequently expressed as combinations of units, like 8 feet 3 inches, may also contribute to students' greater success in converting these units.

Table 7.1

Knowledge of Equivalents of English Units

		Percent Correct	
	Exercise	Age 13	Age 17
1.	Pints in a quart	48	59
2.	Quarts in a gallon	64	77
3.	Ounces in a pound	62	78
4.	Feet in a yard	65	75
5.	Inches in a foot	87	92

The two problems summarized in Table 7.2 indicate that many students who knew basic unit relationships were unable to apply this knowledge to solve simple conversion problems. A comparison of the results for these two problems reveals that two-step problems were more difficult than related problems involving a single operation. It appears that an element of problem solving is involved in even the most straightforward conversion problems, and knowledge of unit equivalents is insufficient to ensure success in solving such problems.

Table 7.2

Conversion Problems

		Percent Correct	
	Exercise	Age 9	Age 13
1.	A piece of board is 60 inches wide. What is its WIDTH in feet?	11	61
2.	Mr. Hernandez needs a ribbon 6 feet 5 inches long. How many inches of ribbon does he need?	--	47

Metric Units

The survey data summarized in Table 7.3 indicate that most students at all age levels have used metric measurement infrequently, if at all. It is possible, however, that they would respond that they used any measurement system only occasionally.

Table 7.3

Use of Metric Units

The metric system of measurement uses units like centimeters, liters, and kilograms. How often have you used the metric system of measurement?

	Percent Responding		
Response	Age 9	Age 13	Age 17
Often	13	23	18
Seldom	46	59	59
Never	27	14	20

The results summarized in Table 7.4 indicate that by the age of 13 most students have some knowledge of common metric units. Although 72 percent of the 17-year-olds could identify which English unit was closest in length to a common metric unit, most students had difficulty thinking of measures in metric units. Only 20 percent of the 9-year-olds, 37 percent of the 13-year-olds, and 50 percent of the 17-year-olds were able to make even reasonable estimates of weight and length in metric units. At each age, these percentages are about 35 points below the performance level for corresponding problems involving English units. Thus, although it appears that most students have had some exposure to metric units, most of them have not learned to "think metric" for even the simplest problems.

Table 7.4

Selection of Appropriate Metric Units

		Percent Correct		
Exercise		Age 9	Age 13	Age 17
1.	Identifies gram as the basic unit of weight*	37	89	92
2.	Which unit would you use to measure the length of your thumb? centimeters meters kilometers	65	80	88

*Unreleased exercise

One of the advantages of the metric system is the ease of converting one metric unit to another. Consequently, one might expect that students who know metric equivalents would have less difficulty applying this knowledge to conversion problems. To some degree, the results summarized in Table 7.5 support this hypothesis. Although results were generally lower for metric problems than for corresponding problems involving English units, a higher proportion of students who knew basic metric relationships could apply them to a simple conversion problem than was true for English units. The performance of 13-year-olds in converting 5 meters 34 centimeters to centimeters compares favorably with their performance on the problem requiring them to convert 6 feet 5 inches to inches. Forty percent of the 13-year-olds correctly answered the former question and 48 percent the latter.

Table 7.5

Knowledge of Metric Relationships

		Percent Correct	
Exercise	Age 9	Age 13	Age 17
1. Identify longest unit of length	--	63	69
2. Centimeters in a meter	--	60	62
3. Millimeters in a meter	--	56	60
4. Meters in a kilometer	--	49	52
5. How many centimeters are 5 meters 34 centimeters?	28	40	52

Change

The greatest improvement shown for any exercises in the entire assessment occurred on the exercises assessing familiarity with common metric units. On the one problem given to 13-year-olds, there was a gain of 26 percentage points; for the two exercises given to 17-year-olds, there was a gain of 12 and 14 percentage points. Apparently the increased emphasis on metric units, both in and out of school, has had some impact, at least in terms of familiarizing students with common metric units.

However, performance on exercises dealing with conversion of English units was uniformly lower for the 1977-78 assessment than it was for the 1972-73 assessment. For both 9-year-olds and 13-year-olds, performance declined about 3 percentage points for most problems, whether they involved simple conversions or more complex applications. It may be inappropriate to attribute this decline to an increased emphasis in the metric system and a corresponding de-emphasis of English units. Performance on an exercise that asked whether 1 month or 5 weeks is a longer period fo time showed a similar decline, despite the fact that months and weeks have not been replaced by metric units.

Performance of 17-year-olds showed little or no change on exercises testing knowledge of basic unit relationships. However, on the three exercises in which they were asked to apply this knowledge, performance declined 8 to 12 percentage points.

Summary

Most students knew the relationships among standard units of measure but had difficulty applying this knowledge to solve conversion problems. Most 13-year-olds and 17-year-olds and about two-thirds of the 9-year-olds have had some exposure to metric measurement. Of these, about 20 percent of the 9-year-olds, 40 percent of the 13-year-olds, and 50 percent of the 17-year-olds appear to be reasonably functional with basic measurement.

Measuring and Estimating Length

Overview of Results

Students were asked to make linear measurements in inches, centimeters, and nonstandard units. To accommodate the testing format, only problems that could be represented on the test booklet were included, and the measurement of longer distances was not tested. On some exercises students were given a twelve-inch ruler marked in sixteenths of an inch, on some they were given a metric ruler marked in centimeters and millimeters, and for others, rulers were printed in the test booklet. Linear estimation skills were also assessed.

Applications of linear measurement skills were tested using maps and scale drawings. The following conclusions can be drawn:

1. Most students could successfully make simple linear measurements, although some students encountered difficulty with fractions of units or measuring the total length of nonlinear paths.
2. Most students have not mastered basic skills in estimating length.

Measuring Length

At all three ages, most students could successfully make simple measurements of length. Eighty-one percent of the 9-year-olds and over 90 percent of the 13- and 17-year-olds could measure the length of a segment that measured a whole number of inches (Table 7.6, exercise 1). Measurements involving fractions of units were significantly more difficult, especially for the 13-year-olds. Only 53 percent of the 13-year-olds and 81 percent of the 17-year-olds correctly identified the length of the pencil in Figure 7.1 to the nearest quarter of an inch (Table 7.6, exercise 3). Even when they were asked to measure to the nearest unit, measurements involving fractions of units caused difficulty for a number of 13-year-olds. This is illustrated by the difference in difficulty of exercise 1 and exercise 2 in Table 7.6.

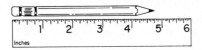

What is the length of this pencil to the nearest quarter inch?
(Note: Rulers were drawn to scale in exercise booklet.)

Figure 7.1. Length measurement exercise.

Table 7.6

Length Measurement Exercises

| | | Percent Correct | |
Exercise	Age 9	Age 13	Age 17
1. Measure to the nearest inch.*	81	91	94
2. Measure to the nearest inch.*	77	77	--
3. Measure to the nearest quarter inch.	--	53	81
4. Measure to the nearest centimeter.*	68	--	--
5. Draw a segment a given number of centimeters.	67	74	--
6. Measure the perimeter of a triangle.	38	65	71

*Unreleased exercise

There was very little difference in respondents' performance on exercises that required them to manipulate a ruler to measure a given length and problems in which a ruler was printed on the page (as in Figure 7.1). There was also very little difference in students' ability to measure given segments and their ability to construct segments of a given length (Table 7.6, exercises 4 and 5).

In general, performance on exercises requiring respondents to measure in centimeters was about 10 percent below performance on similar exercises requiring measurement in inches. However, students did not appear to have great difficulty measuring with arbitrary units. Sixty-four percent of the 9-year-olds, 85 percent of the 13-year-olds, and 89 percent of the 17-year-olds could measure a segment using a paper clip as the unit, and 89 percent of the

9-year-olds could compare the length of two segments by marking the length of one on the edge of a three-by-five card and comparing this measure to the other segment.

Overall, students were reasonably successful in applying linear measurement skills. As might be expected, performance declined somewhat in more complex measurement situations or in problems involving less familiar units, but on the whole, most respondents seemed to understand the basic concept of using a ruler to find the length of a segment. There is some indication, however, that this understanding may be superficial. Although almost all respondents lined up their ruler with the end of the segment when they were measuring it, a problem similar to the one in Figure 7.2 caused considerable difficulty. Only 19 percent of the 9-year-olds and 59 percent of the 13-year-olds correctly answered this question. Seventy-seven percent of the 9-year-olds and 40 percent of the 13-year-olds gave an answer of 5, completely ignoring the fact that the endpoints are not aligned. This is another example of a problem in which a slight change in context generated a superficial response from many students.

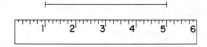

How long is this line segment?*

*This exercise is similar to an unreleased exercise.
(Note: Rulers were drawn to scale in exercise booklet.)

Figure 7.2. Nontraditional measurement exercise.

Students also experienced difficulties measuring more complex figures. Only 38 percent of the 9-year-olds, 65 percent of the 13-year-olds, and 71 percent of the 17-year-olds correctly measured the "distance around" a triangle.

Application of linear measurement skills was assessed using several problems involving scales on maps. One of the exercises is presented in Figure 7.3. Twenty-five percent of the 9-year-olds and 67 percent of the 13-year-olds correctly answered this question.

In the map below, one inch represents four miles. How far is it from Rino to Cott? Use the ruler you have been given to find the answer. (Figure measured 2 inches in the exercise booklet.)

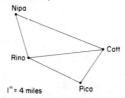

Figure 7.3. Linear measurement application exercise.

No linear measurement exercises from the first assessment were repeated in the second assessment. But comparing current results to results on similar items from the earlier assessment indicates that there probably has not been a significant change in linear measurement skills over the five-year period.

Estimation

Two quite different levels of estimation skill were assessed. One was tested using a multiple-choice format that required respondents to select the only measure that was even close to the height of a common object. These exercises included choices as different as 1 foot, 6 feet, and 15 feet. Essen-

tially, these exercises tested little more than familiarity with the length of
the unit of measure. The second type of exercise asked respondents to estimate the length of a segment to the nearest inch or centimeter, allowing an
error of one unit. This is a much more difficult problem and requires the application of such basic estimation skills as partitioning the segment into
smaller segments. Both types of exercises were administered with metric and
English units. The results are summarized in Table 7.7.

Table 7.7

Length Estimation Exercises

Exercise	Age 9	Percent Correct Age 13	Age 17
1. Gross estimate of the height of a common object in English units*	53	79	86
2. Gross estimate of the height of a common object in metric units*	20	37	50
3. Estimate the length of a segment to the nearest inch.*	35	--	--
4. Estimate the length of a segment to the nearest centimeter.*	29	30	22

*Unreleased exercise

The results indicate that few students at any level have mastered basic
estimation skills. Although there was a great disparity between the metric
and English unit problems that only required general familiarity with units,
there was relatively little difference between the two problems requiring more
precise estimation. Further, the 17-year-olds' performance was considerably
below that of the 13-year-olds for one metric exercise. This might be attributable to the fact that the emphasis on metric measurement occurred after most
17-year-olds had completed any courses that would be affected. However, 17-
year-olds have generally scored higher than 13-year-olds on exercises involving
metric units. In any case, the reversal is worth noting. There were very few
exercises on the entire assessment in which younger respondents outperformed
older ones, and there were none on the first assessment.

Summary

Students are reasonably successful in applying linear measurement skills.
They have some difficulty with problems involving unfamiliar units or fractions
of units and with problems requiring the measurement of a broken line, but most
students seem to understand the basic process of using a ruler to find the
length of a segment. Basic estimation of length, however, has not been mastered by most students.

Telling Time and Reading Measuring Instruments

Overview of Results

There were two basic types of exercises involving time. One was at the
skill level and tested students' ability to read clocks. These exercises were
primarily administered to 9-year-olds. The second type of exercise was at the
problem-solving level and involved finding the time a certain number of hours
after a given time or the amount of time between two given hours of the day.
Students were also shown pictures of a variety of other measuring instruments

like scales and thermometers and asked to identify the indicated measure.

1. Most 9-year-olds could read a clock, although performance declined with finer gradations of time.
2. Problem situations involving intervals of time were difficult at all ages.
3. Students had no difficulty reading a scale if the gradations on the scale represented a whole unit, but they had a great deal of difficulty if the gradations were not equal to a single unit.

Time

Most 9-year-olds could read a clock, although performance declined with finer gradations of time. Ninety-three percent of the 9-year-olds could tell time on the hour, 86 percent could tell time at 15-minute intervals (e.g., 8:15, 6:45), 69 percent could tell time at 5-minute intervals (e.g., 6:25, 11:55), and 59 percent could tell time at 1-minute intervals (e.g., 2:53). At age 13, 85 percent of the respondents could tell time at 1-minute intervals.

Problem situations involving intervals of time proved to be difficult at all ages. Fewer than a third of the 9-year-olds could find the time eight hours after a given time or find the amount of time between two given hours of the day. Only 25 percent of the 17-year-olds correctly solved the following problem:

> A roast is to be cooked 29 minutes for each pound. If a roast weighing 11 pounds is to be done at 6:00 p.m., what time should it be put in the oven to cook?

There is some evidence that performance on these problems may have declined since the last assessment. Performance of 9-year-olds was down 2 percent on one exercise and 14 percent on another. At age 17, performance declined 7 percent on the one problem measuring change. These results are consistent with an overall decline in problem-solving performance. They are probably more attributable to the decline in problem-solving skills in general than to students' failure to learn basic concepts of time. If these results were a valid measure of students' ability to cope with mathematics in real-life situations, they would paint a very bleak picture. All things considered, however, most cooks do manage to get dinner on the table, and most employers would acknowledge that their employees do not have any difficulty figuring out how long until quitting time. These results probably say a great deal more about students' ability to reconcile their everyday experiences with the symbolic problems they encounter in school than they do about students' ability to deal with real-life problems.

Reading Measuring Instruments

Eighty-one percent of the 9-year-olds could read the weight shown on a bathroom scale. However, on an exercise similar to the problem in Table 7.8, students encountered significant difficulty. Their responses suggest that many of them interpreted each mark on the thermometer as a single unit. This represents a basic misconception regarding the nature of units of measure that is not limited simply to reading thermometers.

Table 7.8

Reading a Thermometer

What temperature is shown on this thermometer?*

Response	Percent Responding Age 9	Percent Responding Age 13	Percent Responding Age 17
○ 59°	5	1	0
○ 64°	27	13	5
● 68°	12	46	62
○ 69°	53	40	32
○ I don't know	4	1	0

*This exercise is similar to an unreleased exercise.

Perimeter, Area, and Volume

Overview of Results

Students' knowledge of perimeter was assessed with a series of related problems involving different contexts and vocabulary. Students' ability to calculate areas of figures and their understanding of the notion of unit covering was assessed. Volume skills were also assessed using both pictures of solids that were partitioned into unit cubes and those that were not. Students' ability to apply these skills was assessed with a variety of applied problems requiring them to calculate such things as the cost of carpeting and the amount of concrete needed to pave a driveway. Following are several general observations from the data:

1. Students often rely on superficial responses like adding the numbers shown on the figure to calculate perimeter.
2. Few 9-year-olds have even basic concepts of area.
3. By age 17, most students have some knowledge of area in terms of unit coverings but have difficulty calculating areas of figures from their linear dimensions.
4. There is a tendency to confuse area and perimeter in more difficult problems.
5. Basic concepts of volume are not well developed at any of the ages assessed.
6. Identifying the appropriate unit appears to be a significant problem in understanding basic concepts of area and volume.

Students' ability to find the perimeter of a simple rectangle was assessed with three problems given to different students that involved the same figure but very different contexts. The results are summarized in Table 7.9. As might be expected, many respondents were unfamiliar with the term perimeter, and performance was significantly higher when they were asked to find the "distance around" the rectangle. Performance was not very high, however, for any of the three problems; many respondents adopted the strategy of adding the two given numbers to find the perimeter.

Table 7.9

Perimeter Exercises

Exercise	Percent Responding	
	Age 9	Age 13
1. What is the DISTANCE ALL THE WAY AROUND this rectangle?		
10 ft. 6 ft.		
● 32 feet	40	69
○ 16 feet	39	12
○ 60 feet	4	13
2. What is the PERIMETER of this rectangle?		
10 ft. 6 ft.		
● 32 feet	8	49
○ 16 feet	66	25
○ 60 feet	3	17
3. Mr. Jones put a rectangular fence all the way around his rectangular garden. The garden is ten feet long and six feet wide. How many feet of fencing did he use?		
● 32 feet	9	31
○ 16 feet	59	38
○ 60 feet	14	21

There is a striking difference in performance between the second exercise, in which the rectangle is pictured, and the third exercise, in which the rectangle is described in a problem context without a picture. Since 92 percent of the 13-year-olds could identify a rectangle, most 13-year-olds should have been able to draw the rectangle described in the problem. Since the results for these two exercises are so different, apparently very few of them thought to draw a picture. It seems that this simple problem-solving strategy is not in the repertoire of many 13-year-olds.

Area

At the most basic level, area is defined as the number of units, usually square units, required to cover a given region exactly. Calculations of area by multiplying various linear dimensions of figures should be based on an understanding that such operations are a shortcut for finding the number of units in a unit covering. Assessment exercises measured both students' understanding of the basic concept of a unit covering and their ability to calculate areas of simple geometric regions. There were also about ten exercises that assessed their ability to apply these skills to problem situations. The results for several basic area exercises are summarized in Table 7.10.

Table 7.10

Basic Area Exercises

| | | Percent Correct | |
Exercise	Age 9	Age 13	Age 17
1. Area of a rectangle partitioned into square units*	28	71	--
2. Area of a rectangle*	4	51	74
3. Area of a square*	--	12	42
4. Area of a parallelogram*	--	--	19
5. Area of a right triangle	--	4	18

*Unreleased exercise

Few 9-year-olds have any knowledge of even basic area concepts. Only 28 percent could find the area of a rectangle that was divided into square units, and only 4 percent could calculate the area of the rectangle from the dimensions of the sides. Since most 9-year-olds are still learning basic multiplication concepts, their low performance on the area calculation exercise is not surprising. But they should have had some exposure to the notion of a unit covering.

By age 13, most students can count the number of units in a unit covering, but there are still a substantial number who have not mastered even this basic concept. Furthermore, although 71 percent of the 13-year-olds could find the area of a rectangle that was divided into square units, only 51 percent could calculate the area of the rectangle from the dimensions of the sides. Few 13-year-olds could find the area of more complex figures. Only 4 percent could find the area of a right triangle, and only 12 percent could find the area of a square given one of its sides.

Most 17-year-olds seemed to understand the concept of a unit covering. Eighty-eight percent correctly identified the area of a figure, even though it involved fractions of units, and 76 percent correctly estimated the area of a kidney-shaped region. Seventy-four percent of the 17-year-olds correctly calculated the area of a rectangle. But only 42 percent could find the area of a square given one of its sides, and only about 20 percent could find the area of a right triangle or a parallelogram.

There was some tendency for 13- and 17-year-olds to confuse concepts of area and perimeter. Twenty-three percent of the 13-year-olds and 12 percent of the 17-year-olds calculated the perimeter of a rectangle instead of its area, and about half of the 13-year-olds and a third of the 17-year-olds attempted to find the perimeter of the square and right triangle (Table 7.11) rather than their areas. Precisely why the confusion of area and perimeter was more pronounced for the triangle and square exercises than for the rectangle exercise

is not clear. It may be because the calculations involved in finding the perimeter of the square and triangle were simply the most convenient way to use
the numbers given in the problems, and the error is primarily a function of
students' very superficial notion of both perimeter and area rather than
simply confusing the two.

Table 7.11

Area of a Right Triangle

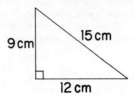

What is the area of this right triangle?

Response	Percent Responding Age 13	Percent Responding Age 17
54 cm^2 or 54	4	18
108 (9 × 12)	4	6
36 (9 + 12 + 15)	48	35
1620 (9 × 12 × 15)	8	4
9, 12, or 15	12	3
Other	13	13
I don't know or no response	11	19

Students' difficulty with area concepts was magnified in their performance on application exercises (Table 7.12). Students even had difficulty
with relatively straightforward problems that required anything more than a
simple calculation. Only 49 percent of the 17-year-olds could find the length
of a rectangle given its area and width, and only 16 percent could find the
area of a region made up of two rectangles.

Table 7.12

Area and Volume Application Exercises

Exercise	Percent Correct Age 17
1. What is the price of a 12 foot by 15 foot piece of carpeting that sells for $7.00 per square yard?	13
2. How many cubic feet of concrete would be needed to pave an area 30 feet long and 20 feet wide with a layer four inches thick?	9

Volume

As might be expected, students' overall knowledge of volume concepts is
poorer than their knowledge of area concepts. Performance on individual exer-

96

cises provides some interesting contrasts that may give insight into the source of students' difficulties. Both parts of the first exercise in Table 7.13 were significantly easier than the second exercise.

Table 7.13

Volume Exercises

---

1. This is one unit:

   What is the volume of each rectangular solid below?

   A.   B.

2. A rectangular solid is cut into cubes as shown. How many cubes are there?

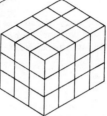

Response	Percent Responding	
	Age 9	Age 13
1A. 4*	76	78
8	3	1
9	5	5
18	1	2
I don't know	8	5
Other	7	9
1B. 12*	35	58
6	26	11
16	15	12
32	1	2
I don't know	8	6
Other	15	11
2. 36*	7	24
33	45	26
66	1	10
I don't know	3	1
Other	44	39

*Indicates correct response

   It is not surprising that part A is easier, since all the cubes are completely visible and can simply be counted. It is worth noting, however, that performance on part A was better than on a similar area exercise in which the units were also visible (Table 7.10, exercise 1). Part B (Table 7.13) does involve some hidden units but was still significantly easier than exercise 2. This may result from the fact that fewer cubes are involved and the problem is easier to visualize. However, it should be noted that in exercise 1, specific attention is called to the unit of volume. Identifying the appropriate unit appears to be a significant problem in understanding area and volume concepts. On exercise 2, 46 percent of the 9-year-olds, 36 percent of the 13-year-olds, and 28 percent of the 17-year-olds simply counted the faces of the cubes shown in the picture or found the surface area of the solid. These students did not understand the meaning of volume and were employing an inappropriate unit of measure. In part B of exercise 1, in which the unit is specified, these errors accounted for only 11 percent and 8 percent, respectively, of the responses of the 9-year-olds and 13-year-olds.

There were also significant differences in the performance of 13-year-olds on the two exercises in which they had to calculate the volume of a rectangular solid by multiplying the lengths of the sides. In one exercise in which the measures were all small one-digit numbers, 33 percent of the 13-year-olds responded correctly. By contrast, another exercise that involved slightly larger numbers was answered correctly by only 17 percent of the 13-year-olds and 39 percent of the 17-year-olds. Both sets of exercises illustrate the difficulty of estimating students' knowledge of a given concept or skill by their performance on a single exercise.

Change

There were only three exercises assessing change in performance on perimeter, area, and volume problems. Performance declined about 4 percent on all three exercises at the 17-year-old level.

Summary

Performance on perimeter, area, and volume exercises was among the poorest of any content area on the assessment. Not only was performance extremely low on exercises at the application level but many students at all ages appeared to have no understanding of the most basic concepts of perimeter, area, and volume.

8

Other Topics

This chapter will discuss results on exercises that dealt with mathematical content that did not fit naturally within the topical headings of other chapters. Probability and statistics, including the organization, description, and presentation of data, are the two major content areas that will be described here. Specific attention was also given to logic exercises, and some discussion of these results will also be included.

Probability

Probability is an important mathematical topic that is included in all current lists of basic skills of mathematics. Sections on probability are found in nearly all mathematics textbooks series, but probability is treated as a luxury topic in many mathematics programs and is often either omitted or included only if time permits.

The second mathematics assessment included twenty-three exercises that addressed various dimensions of probability. Consequently, this assessment gave much more attention to probability than is usually found in any national achievement test. This was done to provide a broad exercise base to document current performance and allow for more monitoring of change in future assessments.

Overview of Results

An examination of the results on these exercises suggests the following key findings:

1. Arrangements involving permutations or combinations were very difficult (fewer than 30 percent correct on any exercise) for 17-year-olds.
2. Calculations and determination of specific probabilities yielded wide ranges of performance for both 13- and 17-year-olds. For example, performance by 13-year-olds ranged from a low of 3 percent correct on one exercise to a high of 84 percent on another, whereas 17-year-olds ranged from 5 to 86 percent correct.
3. Students' intuition had an impact on answers given by all age groups and often resulted in errors. This was most noticeable when calculating the probabilities of independent events and determining expected values.

Permutations and Combinations

Determining the number of arrangements or ordering through the use of permutations and combinations is encountered in many different situations. Specific techniques for determining possible orderings are almost certain to be included within any study of probability. Many problems requiring the computation of the number of different orderings can be determined by simple counting strategies. Indeed, in meaningful learning situations the formulas for finding permutations or combinations are simply extensions of these counting strategies.

Six exercises were related to permutations and combinations. A simple application of permutations is shown in Table 8.1. This exercise could be solved with a counting strategy, and, in fact, this type of ordered-pair problem is frequently used as an application of multiplication. The 13 percent correct by 9-year-olds is far below even the chance level on this multiple-choice exercise and suggests that this application of multiplication was not well understood by the 9-year-olds, even though multiplication is typically studied at this age. Sixty-eight percent of the 13-year-olds answered this exercise correctly, but considering the nature of this problem a much higher level of performance would be expected.

Table 8.1

Results on a Cartesian Product Permutation-Combination Exercise

| Exercise | Percent Responding | |
	Age 9	Age 13

Dale had dark slacks and white slacks.

Dale also had 3 different shirts.

They were plaid, plain, and striped. How many different outfits (slacks and shirts) could Dale wear?

○ 2	59	16
○ 3	9	6
○ 5	16	10
● 6	13	68
○ I don't know	2	0

The two exercises reported in Table 8.2 both involved combinations and could be considered natural extensions of the Cartesian product model illustrated in Table 8.1. Although these exercises, which were given only to 17-year-olds, did not require much calculation, they were very difficult for most of the students. The data indicate that many 17-year-olds did not have any procedure or strategy to call on in solving these problems.

Table 8.2 also provides a comparison of performance on these two exercises during the mathematics assessments conducted in 1972-73 and 1977-78. These results confirm that performance on exercises involving combinations was low in the first assessment, and no discernible improvement was found after five years.

Table 8.2

Change Results on Two Open-Ended Exercises Involving Combinations

Exercise	Percent Correct (Age 17)	
	First Assessment	Second Assessment
A combination lock on a trunk has three dials, one showing all 26 letters of the alphabet and the other two showing the 10 digits 0 to 9. Assuming that each combination uses a setting on all three dials, how many different combinations are possible?	20	15
If each of six teams in a bowling league plays every other team ONCE, how many games will be played in all?	4	5

Probability of Events

Simple events. For both 13- and 17-year-olds, students' grasp of the probability of a simple event was tenuous and seemed to rest heavily on the problem context. For example, the open-ended exercise in Table 8.3 lists the sample space (numbers on the Ping-Pong balls) and states the desired event (finding the ball with a 4). Nearly two-thirds of 13-year-olds and half the 17-year-olds answered incorrectly. Additional analysis of students' responses showed that many of the errors of older students occurred because they were confused about how to report a probability. About 20 percent of the 17-year-olds reported odds (such as 1-5, 1 to 5, 1:5). These errors suggest that many students have some general concept of probability but do not know conventional means of reporting probabilities.

Table 8.3

Change Results on an Open-Ended Probability
Exercise with Sample Space Provided

Exercise	Percent Correct	
	Age 13	Age 17
2, 3, 4, 4, 5, 6, 8, 8, 9, 10		
For a party game each number shown above was painted on a different Ping-Pong ball, and the balls were thoroughly mixed up in a bowl. If a ball is picked from the bowl by a blindfolded person, what is the probability that the ball will have a 4 on it? First Assessment	25	40
Second Assessment	28	35

Students' grasp of the concept of determining the probability of an event seemed to vary depending on the format of the exercise and its context. The exercise in Table 8.3 was an open-ended exercise in an unfamiliar setting, and performance was low. However, when the probability of a simple event was asked with a multiple-choice format, the performance for 13- and 17-year-olds was somewhat higher. On an unreleased multiple-choice exercise, 55 percent of the 13-year-olds and 75 percent of the 17-year-olds responded correctly. All of the foils for this unreleased exercise provided fractions with the same denominator, and so many of the types of errors that occurred in the exercise in Table 8.3 were eliminated. Clearly this is a less demanding task than generating an answer to an open-ended exercise. Unfortunately, it is not possible to determine the effect of exercise format as opposed to problem context. But the results of these exercises suggest that for 13- and 17-year-olds a prime difficulty was reporting the probability of a simple event in standard notation. On exercises that assessed the most elementary concept of probability, however, there were a substantial number of errors that could not be attributed to inappropriate notation.

An unreleased exercise involving the concept of expected value provides some additional insight into students' informal notions of the probability of a simple event. Students were asked how often a coin was likely to come up heads if it were tossed a given number of times. All of the multiple-choice responses were whole numbers, and only one of them was approximately half the number of tosses. About half the 9-year-olds, two-thirds of the 13-year-olds, and three-fourths of the 17-year-olds responded correctly. This exercise and the preceding ones provide a somewhat less pessimistic view of students' knowledge of simple probability concepts than the open-ended exercise in which students had to know the appropriate form for expressing a probability. These three exercises suggest that between one-half and two-thirds of the 13-year-olds and about three-fourths of the 17-year-olds have some informal knowledge of the most elementary probability concepts, but many of them have not had sufficient formal instruction to express probabilities as fractions. As the results of several of the exercises reported below indicate, this limits their ability to deal with more advanced probability concepts.

Closely related to the probability of a simple event is the probability that the event will not occur. Students should recognize that the sum of the probability of an event's occurring and the probability of its not occurring is 1. The probability of an event's not occurring can be found by calculating the probability of all other possible outcomes or by subtracting the probability of the event from 1. Calculations of the probability of an event's not occurring proved to be somewhat more difficult than calculating the probability of an event. In the unreleased multiple-choice exercise mentioned above, 55 percent of the 13-year-olds and 75 percent of the 17-year-olds could find the probability of a simple event. At both age levels, about 15 percent fewer students could find the probability of the event's not occurring.

Compound events. Students' intuitive procedures for dealing with probability caused difficulty with compound events. The context of the exercise in Table 8.4 involving two coins is familiar to most students. There was a marked difference in performance on the two parts of the exercise. In part A, performance was below the chance level for 13-year-olds; part A was correctly answered by fewer than one-third of the 17-year-olds. A majority of both age groups answered part B correctly, even though the processes involved to solve both parts are very similar. What accounts for these differences? A plausible explanation rests on students' reliance on 1/2 as being correct. On each part, a majority of each age group chose 1/2. The percent choosing this response, which happened to be right in one case and wrong in the other, is very similar.

Performance was even lower on an unreleased exercise that involved more than two coins. Only 5 percent of the 17-year-olds correctly calculated the probability of tails turning up on all the coins. Thirty-five percent responded with a fraction with the number of tosses in the denominator. In other words, if there had been four tosses, they would have responded 1/4, or 1 out of 4.

Table 8.4

Results on a Multiple-Choice Exercise in
Which the Sample Space Must Be Formulated

Exercise	Percent Responding	
	Age 13	Age 17
A. Mike has 2 quarters. What is the probability he will get 2 heads when he flips them?		
● 1/4	18	33
○ 1/3	5	3
○ 1/2	58	50
○ 2/3	8	4
○ 3/4	3	2
○ I don't know	8	8
B. What is the probability he will get one head and one tail when he flips them?		
○ 1/4	11	12
○ 1/3	6	3
● 1/2	60	69
○ 2/3	7	4
○ 3/4	8	6
○ I don't know	8	6

Independent Events

Another important probability concept is the notion of independent events. A basic misunderstanding of probability that is often attributed to students and adults alike is the failure to recognize the independence of certain events. If an event has occurred a number of times in a row, it is presumed that the law of averages makes the event unlikely to occur again. Several exercises assessed students' understanding of independent events.

In one exercise, students were told that an event had occurred several times in a row and were asked to find the probability of the event's occurring again. For both 13- and 17-year-olds, performance was about 10 percentage points below that recorded for calculating the probability of the event without the added complication of independent events. Analysis of the incorrect responses for both the simple probability problem and the independent event problem suggested that relatively few students thought the event was less likely to occur because it had already occurred several times in a row. In fact, the most common error was to select the fraction whose numerator was the same as the number of times the event had occurred in the problem. This is consistent with a general pattern of student errors on problem-solving exercises. Many students attempted to use whatever numbers were given in the problem in their response.

An understanding of independent events was not demonstrated by a majority of 13- and 17-year-olds on an unreleased open-ended exercise. Table 8.5 shows that only about one-fifth of the 13-year-olds and two-fifths of the 17-year-olds answered this open-ended question correctly. When the same exercise was given in a multiple-choice format, the performance for 13- and 17-year-olds was 44 and 65 percent correct, respectively. Clearly, the concept of independence of events has not been established for many of these students.

Table 8.5

Results on an Open-Ended Exercise Involving Independent Events

Exercise	Percent Correct Age 13	Percent Correct Age 17
Suppose four consecutive sixes have occurred on four rolls of a fair die. What is the probability of getting a six on the next roll?*	21	39

*This exercise is similar to an unreleased exercise.

The unreleased exercise summarized in Table 8.6 provides some additional insights regarding students' understanding of independent events. Most students did not respond that the long string of girls would make it almost certain that a boy would be born next time. In fact, many students did not pay attention to the information given in the problem--neither the valid information that provided a basis for determining the probability 'nor the information regarding past events that did not affect the probability. Many students simply responded on the basis of their belief that there is an equal chance of a boy or girl being born. Thus, the real difficulty appears not to be with independent events but with deciding what information is relevant for determining a probability and recognizing that one's intuitive notions do not always conform to the real probabilities.

Table 8.6

Results on a Multiple-Choice Exercise Involving Independent Events

Exercise	Percent Responding Age 17
In the United States, of every 1000 babies born, 515 are boys. In a certain U.S. hospital, the last 27 babies born have been girls.	
The next baby born in that hospital will:	
○ almost certainly be a girl (over 80% chance)	4
○ almost certainly be a boy (over 80% chance)	14
● have a slightly greater chance of being a boy than a girl	30
○ have a slightly greater chance of being a girl than a boy	8
○ have an equal chance of being a girl or boy	38
○ I don't know	5

Expected Values

Several exercises involved decision making based on different probability situations, such as expected value. These exercises applied important probability concepts but did not require the calculation of probabilities for specific events. Table 8.7 gives the results on one exercise that provided some data and then asked students to determine the most likely probability model that would produce these results. Forty-one percent of the 13-year-olds and 64 percent of the 17-year-olds responded correctly. Another view of the concept of expected value was provided by an unreleased exercise that was a two-part question in a game format, similar to this:

> Suppose a die is rolled. If an even number appears, you win $6, and if an odd number appears, you lose $4. Will you be more likely to win or lose money if you roll the die once? 100 times?

Most respondents treated these two problems as identical and did not realize that the longer this game is played the greater the opportunity for winning money. This is a key issue in expected value and was not grasped by a majority of the older groups.

Table 8.7

Expected Value

Kim spun a spinner 100 times and made a record of her results.

OUTCOME	A	B	C
NUMBER OF TIMES	55	30	15

Which spinner is most likely the one Kim used? Fill in the oval beside the one you choose.

	Percent Responding	
	Age 13	Age 17
	15	7
	21	12
	16	10
	41	64
I don't know	7	7

105

Summary

All the exercises related to probability were presented through verbal
problems in a realistic context and required very little computation. Most
were given only to 13- and 17-year-olds.

The performance on individual exercises ranged greatly, but performance
of 80 percent or above was found on only five of twenty-five questions given
to 17-year-olds, on only four of twenty-four questions given to 13-year-olds,
and was not reached on any of the eleven questions given to 9-year-olds. All
age groups did better on questions that involved intuitive notions of probability. Much lower performance was found in calculating specific probabilities and issues related to expected value.

All three age groups were, for the most part, very naive about probability. Increases in performance did occur from the 9- to 17-year-olds, but the
overall low performance levels of 17-year-olds raises serious questions regarding their knowledge of basic concepts of probability. Further, things
have not changed appreciably from the 1972-73 assessment. In fact, the performance on change exercises that were used in both assessments showed significant change on only one exercise, and that was a decrease of 5 percent.

Statistics

Statistics is recognized as an area of basic mathematical skill that encompasses a wide number of topics. It is generally recognized that descriptive statistics, such as measures of central tendency and variability, are
useful and provide essential information. Organizing and describing information is also an important component of statistics that relies heavily on experiences with tables and graphs. The assessment included exercises that
dealt with several statistical topics; this section will begin with an examination of student performance on charts, graphs, and tables, and will conclude with an overview of the findings related to descriptive statistics.

Overview of Results

An examination of the statistics exercises revealed two consistent findings:

1. Performance requiring direct reading or comparisons from charts,
    graphs, and tables was consistently higher for all age groups than
    on exercises requiring problem solving.
2. On exercises related to measures of central tendency, overall per-
    formance for all age groups was low. Technical statistical terms,
    such as mean, were found to be unfamiliar to the majority of 13- and
    17-year-olds, although most students knew how to compute a mean.

Charts, Graphs, and Tables

Processing information takes various forms. Reading and interpreting information in graphs and tables represent important information-processing
skills. Such information processing is experienced regularly in reading news-
papers and magazines and in watching television. These presentations take a
variety of forms, such as bar graphs, line graphs, picture graphs, circle
graphs, and tables of varying degrees of complexity.

This assessment contained eighteen exercises that dealt with charts,
graphs, and tables and included a variety of exercises for each age level.
In order to better understand the nature of the exercises, the corresponding
questions related to graphs and tables were classified into four types:

1. Comparisons: Which candidate had the most votes? Who is the short-
    est?
2. Direct reading: How many people attended the game? What is the
    high temperature for Saturday?

3. Interpolation/extrapolation: If brakes are applied at 45 miles per hour, about how far will a vehicle travel after the brakes are applied? (Must estimate from other values reported on graph.)
4. Problem solving: This section includes problems, often multistep problems, that require that information from graphs and tables be used. Some computation is typically required.

An effort was made to examine how the cognitive level of questions influenced overall performance. Table 8.8 reports the range of performance levels of each age group by type of questions asked for all of the exercises related to graphs and tables.

Table 8.8

Percentage Range of Performance Levels by Age
Group and Types of Table Reading Questions

Types of Questions	Age 9	Age 13	Age 17
Direct Reading/Comparisons	30-93	46-98	65-98
Interpolation/Extrapolation	---	5-54	27-72
Problem Solving	29-55	20-87	34-95

For example, the performance by 9-year-olds on direct reading/comparison exercises ranged from a low of 30 percent correct on one exercise to 93 percent correct on another exercise of the same type. The same exercises were not taken by all age groups, and so direct comparisons of general performance levels (i.e., comparing extremes of the age groups) is impossible. Furthermore, direct reading and comparisons were combined in this table, since the ranges of performance were quite similar and the questions were not markedly different. These ranges show that asking students to do something beyond simply reading data resulted in dramatic changes in performance. Table 8.8 also documents that the lowest performance for both 13- and 17-year-olds resulted from tasks requiring them to interpolate or extrapolate from data that were given.

On all graphical data, questions involving direct reading and comparisons were easier for each age group than questions requiring interpolation/extrapolation or problem solving. This is clearly illustrated by the exercise in Table 8.9, which was given only to 17-year-olds. The table shows that 87 percent responded correctly on a question requiring direct reading of a graph, whereas performance dropped to 66 and 36 percent on questions requiring some problem solving as well as computation.

Although students could read graphs and tables, they may not be as successful at constructing graphs. Students were not asked to draw a graph, but 9-year-olds were given data and asked to select the graph that represented the information (Table 8.10). The 29 percent correct was only slightly above the chance level for this multiple-choice exercise; this confirms that this type of problem solving was very difficult.

Several different exercises were used to measure change in 13- and 17-year-olds' performance on graphs and tables from the 1972-73 assessment to 1977-78. None of the change exercises are released, but a typical exercise presented a sales-tax chart to 13- and 17-year-olds. Several prices requiring direct reading from this chart were given, and respondents were asked to find the appropriate tax. On four separate questions, the performance of 13-year-olds was between 46 and 49 percent correct (an average drop of 4.5 percent), while that of the 17-year-olds ranged from 75 to 79 percent correct (an average drop of 3.5 percent). This type of small but consistent decline was observed on all but one of the graph and table exercises given to measure change in performance.

Table 8.9

Results on Reading and Using a Graph

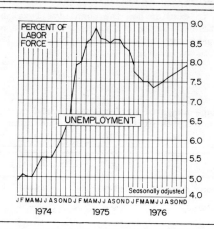

	Percent Responding Age 17
A. What percent of the labor force was unemployed in July 1974?	
○ 4.9	4
● 5.5	87
○ 7.5	2
○ 8.6	4
○ I don't know	3
B. What was the change in percent unemployed from July 1974 to July 1976?	
○ 0.0	0
● 2.0	66
○ 3.1	9
○ 5.5	4
○ 7.5	15
○ I don't know	6
C. If the labor force totaled 98 million in July 1974, approximately how many employable Americans were out of work that month?	
○ 4.80 million	16
● 5.39 million	36
○ 7.35 million	8
○ 8.43 million	8
○ I don't know	32

Table 8.10

Nonroutine Graphing

Here are the ages of six boys:

10, 9, 9, 8, 8, 8

Which bar graph presents this information?

	Percent Responding Age 9

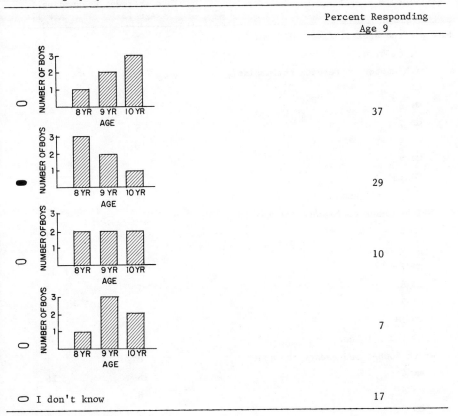

○	37
●	29
○	10
○	7
○ I don't know	17

Descriptive Statistics

The bulk of the NAEP exercises dealing with statistics related to averages, including more technical terms such as mean, median, and mode. Most of these exercises were administered to 13- and 17-year-olds, since few elementary school students have much experience with averages in their mathematics classes. Averages are usually encountered in junior and senior high school mathematics programs, and so this assessment reflects the typical emphasis in most mathematics curricula.

Measures of central tendency—namely, median, mean, and mode—were difficult for both 13- and 17-year-olds, as illustrated by the three-part exercise reported in Table 8.11. In no case did a majority of the respondents answer any of the questions correctly, and in fact, the performance levels of both 13- and 17-year-olds were nearly the same. For example, the median was computed correctly by 42 percent of the 13-year-olds and 43 percent of the 17-year-olds; the mean by 6 and 12 percent, respectively, and the mode by 25 and 26 percent, respectively. The large percentage of "I don't know" responses, reaching as high as 36 percent for the 17-year-olds, suggests that these terms are unfamiliar to many students.

Table 8.11

Results on Exercises on Measures of Central Tendency

Exercise	Percent Responding	
	Age 13	Age 17
Here are the ages of five children:		
13, 8, 6, 4, 4		
A. Which number represents the median?		
○ 4	13	8
● 6	42	43
○ 7	9	19
○ 8	10	11
○ 9	1	3
○ 13	9	5
○ I don't know	16	11
B. Which number represents the mean?		
○ 4	18	21
○ 6	4	6
● 7	6	12
○ 8	7	5
○ 9	2	2
○ 13	32	29
○ I don't know	31	25
C. Which number represents the mode?		
● 4	25	26
○ 6	8	5
○ 7	4	4
○ 8	9	6
○ 9	3	2
○ 13	17	21
○ I don't know	34	36

An unreleased exercise presented the same data but used the term _average_ rather than _mean_. This one-word change in these exercises generated two interesting results. First, the "I don't know" responses dropped from 31 to 9 percent for the 13-year-olds and from 25 to 3 percent for the 17-year-olds. Second, the performance for 13-year-olds jumped from 6 to 51 percent and from 12 to 71 percent for 17-year-olds. These changes are dramatic and show that a majority of the students are more familiar with the term _average_ and interpret average as the arithmetic mean. Furthermore, a majority of the students can correctly compute an arithmetic mean when they know what they are looking for. These results once again document the impact that terminology can have on performance.

A different application was required in the exercise reported in Table 8.12. Instead of computing statistics, the exercise asked 13-year-olds to select appropriate statistics for a real-world application. Two questions, one involving the mode and the other involving the range, were included. Fewer than half of the students answered either question correctly, with only 12 percent correctly choosing the mode in a real-life situation that required its use. This performance level is about half that reached when a direct computation of the mode was requested, as in Table 8.11, part C. Thus, the selection of the proper statistic in this application context was much more difficult than simply computing the statistic.

Table 8.12

Results on an Application of Measures of Central Tendency

Exercise	Percent Responding Age 13
A. If you could reorder <u>only one shoe size</u> for your shoe store, which statistic would be most helpful?	
● The mode of the sizes sold recently	12
○ The mean of the sizes sold recently	10
○ The median of the sizes sold recently	13
○ Either the mean or the median of the sizes sold recently	15
○ The range of the sizes sold recently	29
○ I don't know	21
B. If you wanted to know how many different shoe sizes to order, which statistic would be most helpful?	
○ The mode of the sizes sold recently	13
○ The mean of the sizes sold recently	9
○ The median of the sizes sold recently	9
○ Either the mean or the median of the sizes sold recently	13
● The range of the sizes sold recently	40
○ I don't know	16

Three different exercises give additional information about students' knowledge of average. A brief description of each type is provided in Table 8.13. The first exercise was a typical textbook problem involving average. The other two exercises (Table 8.13, parts B and C) were not complex but did require a greater understanding of average than the other exercise. These exercises required very little computation, but both were multistep problems. In each case the performance dropped from that on the routine problem.

The concept of weighted mean was also included in several exercises. Performance on such exercises was low. For example, a typical exercise might read:

If 4 pounds of ground beef is bought at $2.00 per pound and 5 pounds of round steak at $3.00 per pound, what is the average cost per pound for this meat?

About 7 percent of the 13-year-olds and 17 percent of the 17-year-olds answered a multiple-choice version of this exercise correctly. A similar open-ended exercise was presented only to 17-year-olds, and 13 percent of them answered it correctly.

These performances were low, and a closer examination of these results suggests that an overwhelming majority of 13- and 17-year-olds did not have a good grasp of the concept of weighted average. For example, two of the most frequent errors resulted in unreasonable answers, yet they were offered by many respondents. The most predominant error (made by 20 percent of 13-year-olds and 34 percent of 17-year-olds) was to add the per-pound costs ($2 plus $3) and the weight (4 pounds plus 5 pounds) and divide this cost by the total weight. This approach results in an average ($0.56 per pound) that is far less than the price of either item separately. This answer is clearly unreasonable but was still chosen by more than one-third of the 17-year-olds. Similarly, 30 percent of the 13-year-olds and 10 percent of the 17-year-olds simply added the cost per pound together and reported $5 as the average. In both situations, the fact that this average was not between the given prices apparently did not discourage respondents from reporting those values.

Table 8.13

Summary of Results on Exercises Involving Average

Description of Exercise	Percent Correct		
	Age 9	Age 13	Age 17
A. Given total and numbers of entries--find average.	22	73	83
B. Given total for a specific number of entries--project total after a different number of entries.	*	47	69
C. Given average and values of all but one entry--find value of missing entry for average to be reached.	*	16	26

*Not given to 9-year-olds

Only two of the statistics exercises measured change in performance from the first to the second assessment. No pattern of consistent change was found in these exercises. In fact, the percentage on these exercises was no greater than would be expected due to sampling error.

Logic

Logical reasoning is recognized as an important skill. Many of the exercises throughout the mathematics assessment required the use of logic either directly or indirectly in a variety of different ways. This discussion is based on ten multiple-choice exercises that dealt with some specific logical reasoning tasks.

Overview of Results

An examination of the results on this group of exercises leads to the following observations:

1. The 17-year-olds consistently performed near or above the 80 percent correct level on the simple logic exercises administered.
2. A much larger jump in performance typically occurred between the ages of 9 and 13 (as high as 33 percent) than between ages 13 and 17 (generally around 10 percent).
3. Performance dropped from levels on the first assessment on all logic exercises common to both assessments.

Table 8.14 reports results that illustrate high performance by 17-year-olds, as well as marked differences between the performance levels of 9- and 13-year-olds. The increase in performance from age 9 to 13 is a reflection of the significant cognitive growth that is typically experienced between these ages. Although growth continues into the oldest group, it is not as dramatic.

Table 8.14

Results on a Logic Exercise

Exercise	Percent Responding		
	Age 9	Age 13	Age 17
Two team captains take turns choosing players for their teams. Ellen is always chosen first. Chris is always chosen second. If Ellen and Chris are never captains, how often do they play on the same team?			
◯ Always	26	16	9
◯ Frequently	17	11	6
◯ Very rarely	16	9	6
● Never	33	62	78
◯ I don't know	8	2	1

Several more classical situations involving deductive information were used. Specific conditions were always presented in a real-world setting. For example, an unreleased exercise provided a conditional statement similar to this:

If Kelley catches a fish, then the fish is small.
The fish is big.

Over 80 percent of each age group concluded that Kelley could not have caught the fish. In fact, 96 percent of the 17-year-olds answered the question correctly.

On the other hand, performance for all age groups dropped when they were given a deductive reasoning problem with insufficient information to make a conclusive statement. For example, consider the assertion:

Every flyer is crazy.
Chris is crazy.

A correct conclusion that there is not sufficient information to decide whether Chris is a flyer was reached by 25, 51, and 58 percent, respectively, of the 9-, 13-, and 17-year-olds.

Changes in performance from the first assessment are illustrated in Table 8.15. This exercise was presented in a more symbolic mode than most of the other related logic exercises. Although the significant changes here are greater than those found on the other logic exercises, there were decreases on every logic exercise given in both assessments.

Table 8.15

Results on a Logical Reasoning Exercise

| Exercise | Age | Percent Correct | |
If a > 5 and b > 5, then		First Assessment	Second Assessment
⊃ a equals b	9	*	27
⊃ a is greater than b	13	67	57
⊃ b is greater than a	17	69	76
● there is not enough information to determine the relation between a and b			
⊃ I don't know			

*Not administered to 9-year-olds in the first assessment

Despite the fact that performance on change exercises dropped, the generally high level of performance of 17-year-olds on logic exercises is encouraging. It is also a bit surprising, particularly when the amount of instruction devoted to developing logical reasoning is considered. Can it be that such skills are developed naturally, apart from the mathematics classroom?

The dramatic increases in performance from age 9 to age 13 are consistent with learning theory that identifies this period as a critical time in developing formal reasoning. It would be interesting to know what portion of these changes is attributable to the school curriculum and what part is simply the result of mental maturity. The results do not provide the answer, but the changes observed from age 13 to age 17 were much more modest.

9

Calculators and Computers

Calculators

The 1977-78 mathematics assessment offered a unique opportunity to gather information that will help clarify how students performed with calculators and will help shape the direction of future research in the area. These benchmark data will be valuable as the availability of hand calculators and their impact on mathematics programs are studied during the 1980s.

Calculators are available to students. Seventy-five percent of the 9-year-olds, 80 percent of the 13-year-olds, and 85 percent of the 17-year-olds in the assessment stated that they had access to at least one calculator. In addition to determining the availability of calculators, this assessment attempted to ascertain how well students solved different types of mathematics problems when they were familiarized with a simple, four-function calculator.

NAEP Procedures for Calculator Exercises

To assess students' ability to use calculators, a special calculator booklet was prepared. Students at each age level were presented with a typical NAEP test booklet. After the standard instructions were given on how to answer both multiple-choice and open-ended exercises, students who participated in the calculator portion of the assessment were given brief instructions on the operation of a TI-1200 hand calculator. These instructions included turning on the calculator, clearing the display, entering a number that included a decimal, and working a simple addition and subtraction problem. The students were also given several minutes to work other problems for which they knew the answer. These instructions were included on a single page of the test booklet and extended only to showing students how to use this particular model of calculator. No attempt was made to acquaint them with special keys, such as the percent key, or to teach them how to use a calculator for the first time. Students were instructed to use the calculator on all exercises within the booklet; use was not intended to be optional.

Approximately twenty-five exercises were administered at each age level, with many exercises given to more than one age group. In all, forty-seven different exercises were included in this portion of the assessment, sixteen of which have been released. Many of the same exercises were presented to other students in a paper/pencil format as part of the regular assessment. Timing for each format was identical. It should be noted that many test administrators reported that some students took longer to do problems with a calculator than with paper and pencil. Nevertheless, the timing allowance for the paper/pencil and calculator exercises was consistently adhered to throughout the testing of all age groups.

Categories Assessed

Four categories of exercises were included in the hand calculator package of the assessment:

1. Routine computation. This category included problems for which students had been taught the appropriate paper-and-pencil algorithm.

2. Nonroutine computation. This category included problems for which students had not formally been taught an algorithm.
3. Concepts and understanding. This category included such topics as order of magnitude of fractions and decimals, approximating square roots, and concepts related to percent.
4. Applications and problem solving. This category included routine word problems as well as more difficult multistep problems. For 13- and 17-year-olds, emphasis was on consumer problems.

Routine Computation

Overview of Results

Comparing performance with and without the calculator for different age groups on the routine computation exercises included in this assessment revealed the following key findings:

1. Performance for 9-year-olds on simple addition exercises remained relatively consistent whether done with or without the aid of a calculator. Slight increases in performance on addition exercises were shown by 13- and 17-year-olds when a calculator was available.
2. Performance levels by all age groups on subtraction, multiplication, and division with whole numbers showed increases when a calculator was used.
3. Nine-year-olds' performance on subtraction, multiplication, and division computation with a calculator was only slightly lower than that of 13- and 17-year-olds without a calculator.
4. Thirteen- and 17-year-olds performed better on exercises involving decimals with the aid of a calculator than they did without a calculator.
5. The format for division exercises affected how all age groups used the calculator. More specifically, errors were frequently made through reversing the dividend and divisor in the keystroking sequence.
6. When using calculators, students of all age groups were more likely to make no response to exercises than they were on noncalculator exercises. At the same time, fewer students responded "I don't know" when using calculators.

Discussion of Results

Thirteen exercises, each including multiple parts, were constructed for this portion of the hand calculator assessment; some of these that are representative are described and the performance levels given in Table 9.1. The exercises included a sampling of each of the four fundamental operations in a variety of formats and included from three to twelve digits. The exercises differed from those in the rest of the calculator assessment in that no decision was required of the student as to what digits or what operation should be used. The only task expected of each student was to use the calculator to perform the indicated computation.

Parts A, B, and D of the addition exercise in Table 9.1 were given to 9-year-olds in both calculator and paper/pencil booklets. On parts A and D the calculator seemed to make little difference, but on part B calculator performance was 7 percentage points higher than its companion paper/pencil exercise. Part of this increase in performance may have been because B required regrouping in the paper/pencil format, and computation that requires regrouping has traditionally been a stumbling block for many students. With a calculator, however, the act of regrouping is done internally without special input from the operator. Another possible explanation is that part D required the operation symbol (+) to be entered at least twice, a procedure that may have caused some confusion. This conjecture is supported by a 20 percent no-response increase for this exercise.

Table 9.1

Performance on Selected Routine Computation Exercises

Operation	Age 9 NC*	Age 9 C	Age 13 NC	Age 13 C	Age 17 NC	Age 17 C
Addition						
A. 21 +54	90	89	98	--	98	--
B. 37 +18	76	83	95	--	97	--
C. 43 71 75 +92	50	--	84	92	92	97
D. 4285 3273 +5125	51	50	85	--	90	--
Subtraction						
A. 231** -189	50	80	85	95	--	98
B. 4.2 - .67**	--	24***	34	80	63	87
Multiplication						
671 ×402	3	74***	66	92	77	96
Division						
A. 4)76	19	71	--	--	--	--
B. 28)3052	--	50***	46	82	50	91

*NC = No calculator available; C = calculator
**This exercise is similar to an unreleased exercise.
***This exercise represents a nonroutine computation exercise for 9-year-olds.

Generally, however, these representative results show that 9-year-olds'
ability to add was not noticeably affected when a calculator was used. Thir-
teen- and 17-year-olds were given only one exercise (part C) for addition in
both formats. Based on this one exercise, performance was slightly better (6
to 8 percentage points) for each age group when a calculator was available.
The results shown in Table 9.1 also provide evidence that the calculator
aided every age group on the operations of subtraction, multiplication, and
division. Improvement in several cases measured around 50 percentage points.
In all cases the calculator proved an accurate, efficient computational tool,
much more so than paper/pencil calculation. An interesting sidelight to this
finding is the fact that 9-year-olds' computational ability with the calculator
was only slightly below that of 13- and 17-year-olds without a calculator.
These differences never exceeded more than 9 percentage points between 9- and
13-year-olds and never more than 13 points between 9- and 17-year-
olds.
With only two exceptions, 9-year-olds' performance with a calculator declined with an increase in the number of digits involved in the computation.
For example, on division exercises that contained a total of three digits,
around three-fourths of the 9-year-olds responded correctly, but when six

digits were involved, only half of them were correct. As might be expected, the highest performance with calculators on routine computation, as well as the other categories of exercises, was turned in by 17-year-olds, followed by 13-year-olds, and then by 9-year-olds. Older students report greater access to hand calculators than younger students. This is likely to have an effect on both familiarity and efficiency of operation.

Three unreleased decimal exercises--two subtraction exercises and one division exercise--were given to 13- and 17-year-olds in both the calculator and paper/pencil booklets. On two of these exercises the calculator helped each age group raise its performance by as much as 46 percentage points for 13-year-olds and 26 percentage points for 17-year-olds. For both age groups, problems in keystroking the division exercise occurred, with 28 percent of the 13-year-olds and 12 percent of the 17-year-olds reversing the dividend and divisor.

Students experienced a similar difficulty on the third decimal exercise, a subtraction problem, in deciding which number was the subtrahend and which was the minuend. Performance for both age groups showed only slight differences (less than 3 percentage points) between the calculator and paper/pencil groups for this subtraction exercise.

The only decision needed when using a calculator to divide is choosing which number is to be entered first in the keystroking sequence. Although performance on division exercises using the calculator was high for all age groups, this assessment documents the obvious confusion over which number is the dividend. The format in which the division exercise was presented tended to influence the number of reversal errors. Division problems were presented in three formats to 9-year-olds, and as Table 9.2 shows, the format considerably influenced the performance level.

Table 9.2

Nine-Year-Olds' Performance on Three Division
Exercises Presented in Different Formats

	Format	Percent	
		Correct	Reversal Error
A.	$3\overline{)36}$ **	71	8
B.	Divide 7 into 42*	36	37
C.	Divide 72 by 9*	71	5

*Similar to an unreleased exercise

This reversal problem was also demonstrated in other exercises taken by 13- and 17-year-olds (Table 9.3). Obviously the decision as to which number to divide by must be made whether or not a calculator is available. These results provide a vivid reminder of the need to teach division meaningfully, so that the written symbols are well understood.

Table 9.3

Performance on a Decimal Division Exercise

Exercise		Percent			
		Age 13		Age 17	
		NC*	C	NC	C
$.04\overline{)8.4}$ **	Correct	34	55	52	78
	Reversal Error	0	28	0	12

*NC = No calculator available; C = calculator
**Similar to an unreleased exercise

The use of calculators tended to affect the response rate of students of each age group. On almost every exercise presented in both paper/pencil and calculator booklets, the percentage of no response was higher for the calculator booklet. This tendency toward no response was most noticeable on problem-solving/applications exercises and will be discussed in further detail later in that section.

Nonroutine Computation

Overview of Results

Decisions about what tasks are nonroutine for different groups of students may be open to question. For purposes of this discussion, nonroutine computation exercises are those for which no formal algorithm would have been taught for most students in the appropriate age group or for which using a calculator would yield unfamiliar output. When defined in this way, nonroutine computation included exercises involving decimals and long division exercises with two or more digits in the divisor for 9-year-olds, an exercise dealing with integers for 13-year-olds, and a display overload exercise for 17-year-olds. This category also included exercises that dealt with order of magnitude of fractions and decimals, approximating square and cube roots, and computation with percents.

The major results are these:

1. Performance on all nonroutine computation exercises was poor, with no improvement shown when a calculator was available.
2. Students at all age levels experienced difficulties in entering data and interpreting output in unfamiliar situations.

Discussion of Results

Table 9.4 summarizes performance on several nonroutine decimal computation exercises given to 9-year-olds in this calculator assessment. Performance in general was low, and in no instance in which exercises were given in both calculator and paper/pencil booklets did the calculator improve performance.

Each age group had varying degrees of difficulty entering, reading, and interpreting decimals when they appeared within exercises and when they resulted from computation on the calculator. Nine-year-olds tended to ignore decimals. For example, on the unreleased multiplication exercise shown in Table 9.4, 26 percent of the 9-year-olds responded correctly and 61 percent responded with an answer that contained a decimal placement error.

Results on the exercise shown in Table 9.5 further illustrate some of the confusion students felt when dealing with decimals. Nine-year-olds keyed in the sequence of numbers shown in the exercise; however, 26 percent had difficulty transferring the displayed answer to the appropriate answer in the multiple-choice list. Since the numeral 5384615 was the first choice containing the significant digits desired, they may have stopped at this point. This conjecture can be supported by results on part B of this exercise.

Few 9-year-olds reversed the dividend and the divisor. This response was more common for 13- and 17-year-olds. It is surprising that such a high percentage of both 13- and 17-year-olds chose the answer 5384615 to part A. Around 20 percent of both age groups chose this answer over the correct one even though it is clearly unreasonable and despite the fact that the exact answer must have been displayed on their calculator. Clearly, instructing students on how to enter a decimal, as was done in the instructions for this

Table 9.4

Performance on Selected Nonroutine Computation Exercises

Operation	Percent Correct (Age 9) NC*	C
Addition		
A. .35 +1.4	19	14
B. 7.54 +1.52	32	25
Subtraction		
A. 8.47** - .25	--	28
B. 4.2 - .67**	--	24
C. 23.7** - 8.56	--	37
Multiplication		
.6 × 4.3**	--	26
Division		
6)608	--	37

*NC = No calculator available; C = calculator
**Similar to an unreleased exercise

Table 9.5

Performance on a Calculator Division Exercise

Exercise	Percent Responding Age 9	Age 13	Age 17
A. 7 ÷ 13 =			
◯ 18571428	2	7	3
◯ 1.8571428	2	13	10
◯ 5384615	26	22	20
● 0.5384615	51	50	64
◯ I don't know	9	2	1
◯ No response	11	7	3
B. 10 ÷ 25 =			
● 0.4	65	73	84
◯ 2.5	4	17	12
◯ 4	10	4	1
◯ 25	2	1	1
◯ I don't know	7	1	1
◯ No response	12	5	2

portion of the assessment, is not enough to expect any age group to deal appropriately with computation involving either decimal input or output.

A similar problem of being unable to read the calculator display occurred in the exercise shown in Table 9.6, in which 13- and 17-year-olds were asked to compute answers to four equations involving positive and negative integers. It should be noted that students were not provided any instruction on how to enter negative integers on the particular model of calculator used. Furthermore, the TI-1200 has no change sign key for entering negative integers. There is no direct method for computing an equation such as 4 - (-6), which, if keystroked on the TI-1200, results in -2.

Table 9.6

Responses for 13- and 17-Year-Olds
on a Nonroutine Calculator Exercise

Exercise*	Percent			
	Age 13		Age 17	
	Correct	No Response	Correct	No Response
A. $^-127 + {}^+36$	49	3	75	2
B. $^+47 - {}^-33$	12	11	38	8
C. $^-26 \times {}^-14$	49	27	65	20
D. $^-295 \div {}^+40$	14	43	38	39

*Similar to an unreleased exercise

Performance on the exercises shown in Table 9.6 was poor, but because the calculator used was not designed to compute with negative integers, the performance might have been as good as could be expected. If students were unfamiliar with the computation of integers, the only alternative would be to use exact keystroking--that is, keystroking that replicates the equation, using the subtraction key for the negative symbol.

The best performance by either age group was on parts A and C. On part A, unlike parts B and D, if the exact keystroking were used, a correct answer would result. On parts B and D, however, exact keystroking yields an incorrect and, at least for part D, a clearly unreasonable answer. Unless the student understood that an alternative to subtracting a negative integer would be to add its additive inverse, part B could not be successfully completed with the TI-1200.

The high percentage of students making no response on part D is difficult to explain. Perhaps one explanation is that using the exact keystroking provided in the exercise gave an answer students recognized as unreasonable. A more likely explanation is that students did not have adequate time to complete all parts of this exercise.

A clear implication from results of this exercise is that in a testing situation, there is a need to familiarize students with the logic behind the particular calculator they are using as well as the purpose of any special keys. Without this instruction, we may be doing little more than asking students to guess the correct answer.

Two computation exercises given to 17-year-olds to be done with a calculator involved multiplication that would overload the eight-digit display of the TI-1200. When overloaded, the display showed as many as eight digits of the resulting computation but blinked repeatedly to indicate the overload. The first of these exercises resulted in a product with two significant digits and was correctly answered by 13 percent of the 17-year-olds; 40 percent made no response. Only 2 percent said they did not know the answer. The second portion of this exercise also involved a multiplication problem, but this one had a product of nine significant digits, which could not be reported in entirety on the display of the calculator. Only 1 percent of the 17-year-olds correctly identified the answer in this open-ended exercise. Fifty-seven percent made no response, and 2 percent responded "I don't know." Also, 28 percent reported a blinking (error) display.

Clearly, these exercises forced the student to seek strategies for the solution not typical of their mathematics classroom experience. The majority were unable to do this and did not respond to the exercise.

Concepts and Understanding

Overview of Results

This category included exercises that dealt with the following topics: order of magnitude of fractions and decimals, approximating square and cube roots, and understanding percents. In all, five exercises will be discussed in this category. All exercises were administered only to 13- and 17-year-olds. Following are the major observations for which supporting data will be presented:

1. Performance on all exercises included in this portion of the calculator assessment was very poor.
2. Thirteen- and 17-year-olds were unable to use a calculator to order a set of fractions between 0 and 1.
3. The concept of percents is not understood by those 13- and 17-year-olds tested. Consequently, performance on exercises with a calculator was either poorer than, or only slightly above, the performance on the same exercises without a calculator.

Discussion of Results

Two exercises involved ordering fractions (Table 9.7, exercises 1 and 2). It seems reasonable that a calculator would increase the efficiency of this process, but the performance level of 13 percent correct by the 17-year-olds shows that this was not the case. Perhaps an explanation for this low performance is that the process of changing the fractions to decimal form with the calculator was unfamiliar to students, or perhaps students did not realize that the calculator would perform this computation. Whatever the case, performance on this exercise was disappointing.

This lack of understanding was clearly illustrated in several other exercises, and it points to the fact that calculators will only compute and will not help a student decide what to do. This fact is further supported by results of what is perhaps the most difficult problem in this calculator assessment, exercise 3 in Table 9.7, given to 17-year-olds in the calculator format only. In this exercise, at least two situations must be satisfied in order for the student to arrive at a correct solution. First, an understanding of the concept of an exponent is required, and second, an efficient, accurate computational device is needed to allow use of trial-and-error methods for the solution. The calculator satisfied the latter need; however, only 4 percent of the 17-year-olds correctly answered this question, with 38 percent making no response and 7 percent responding "I don't know." Although the calculator should have aided in the solution of this exercise, it obviously did not. Again it can be said that students can perform only to the extent that they understand the concept. A calculator, in this case, did not aid in this understanding.

A similar statement applies to the unreleased open-ended exercise that asked 13- and 17-year-olds to use calculators to find the square root and cube root of a given number (the actual exercise was expressed in the form of $\Box \times \Box = 196$). The calculator had no square root button, and so the students had to find the values by trial and error. Although both answers were whole numbers, fewer than 13 percent of the 13-year-olds could correctly answer either

Table 9.7

Results on Concepts and Understanding
Calculator Exercises

Exercise	Percent Correct			
	Age 13		Age 17	
	NC*	C	NC	C
1. Arrange the following fractions from the SMALLEST to the LARGEST. $\frac{5}{8}, \frac{3}{10}, \frac{3}{5}, \frac{1}{4}, \frac{2}{3}, \frac{1}{2}$	2	2	12	13
2. Choose which of two fractions is closer to 1.	57	52	--	61
3. Find the smallest integer n such that $(.8)^n \leq .6$**	--	--	--	4
4. A. Find square root of given number	--	10	--	37
B. Find cube root of given number	--	12	--	34
5. A. 30 is what percent of 60?	35	25	58	--
B. What is 4% of 75?	8	8	27	36
C. 12 is 15% of what number?	4	3	12	19
D. What is 125% of 40?	12	10	31	24
E. 6 is what percent of 120?	6	6	16	15

*NC = No calculator available; C = calculator
**Similar to an unreleased exercise

question, and fewer than 37 percent of the 17-year-olds correctly found either
solution. For both age groups, a large number of students made no response.
The percentages giving no response were higher among 17-year-olds (28 and 33
percent, respectively, for each portion of the exercise) than for 13-year-olds.

Once again, solution of these exercises by trial and error is much more
efficient with a calculator and should require only a minimum amount of time.
The large number giving no response is consistent with many other exercises
done with a calculator. Either students did not have time to complete their
exercise, or else they did not have an efficient strategy for finding the
solution.

Exercise 5 in Table 9.7 examined student performance on a set of traditional percent problems. The results on this open-ended exercise again illustrate that performance with calculators rarely exceeded the level of performance without calculators. For example, consider part C of exercise 5, which
was given in both calculator and paper/pencil booklets to 13- and 17-year-olds.
Without a calculator, 4 and 12 percent of 13- and 17-year-olds, respectively,
answered correctly. A question that might be asked is whether this low performance reflects a lack of understanding of the concept or whether it reflects
poor computational ability on the part of these age groups. A partial answer
to this question can be given if we look at performance on this same question
where a calculator was provided to aid computation. Three percent and 19 percent, respectively, of the 13- and 17-year-olds with a calculator correctly

answered this same question. It would seem that computation ability accounts for only a small portion of the difficulty, especially for 13-year-olds. Given the already documented difficulty 13-year-olds have dealing with decimals when using the calculator, perhaps this also acounts for a portion of the low performance.

The low performance on these percent exercises is disconcerting, but of more consequence is the fact that computation ability is apparently not the main problem. As evidenced by this exercise, students need a stronger conceptual development of the topic. It will do little good to drill and reinforce computation with percents if students do not understand when or where to compute, or are not able to determine if their answers are reasonable. Also of interest in this exercise is the consistent increase in the number of students making no response at all when the calculator was used with an accompanying decrease in the number of "I don't know" responses for each age group. This is true about each of the five exercises. Reasons for the tendency not to respond are speculative, but it may be that students were reluctant to admit they could not answer these questions when a calculator was available. Perhaps they felt, like many adults, that a calculator should enable them to answer any of these questions.

Problem Solving and Applications

Overview of Results

Every exercise included in this category of the hand calculator assessment included more than simple computation. All such exercises required decisions about what operation(s) to use, and many included decimals, percents, or money. Comparing performance for each age group with and without calculators on these problem-solving/application exercises yields the following results:

1. Nine-year-olds performed better on single-step problems than on multistep problems when a calculator was used. For multistep exercises, 9-year-olds' performance was very low, with the calculator format yielding slightly lower results than the paper/pencil booklet.
2. Thirteen-year-olds using a calculator performed worse on 7 of the 10 exercises in this category than those given the same problems in paper/pencil booklets.
3. Seventeen-year-olds' performance with a calculator was better on 9 of the 14 exercises given in both calculator and paper/pencil formats.
4. The lowest performance for all groups was found on exercises involving division.

Discussion of Results

Eleven single-step exercises were included in this portion of the assessment. All but three were given to 9-year-olds in a calculator booklet alone or in both calculator and paper/pencil booklets. A variety were also given to 13- and 17-year-olds both with and without a calculator.

For 9-year-olds, performance with the calculator was higher than performance on the paper/pencil version for each single-step exercise given in both calculator and paper/pencil booklets. The degree of difference ranged from 6 to 32 percentage points. The two multistep exercises given in both formats to 9-year-olds showed slight decreases (differences of 5 and 4 percentage points) when a calculator was available.

Seven one-step exercises were given to 13-year-olds in both calculator and paper/pencil booklets. These exercises involved decimals, a factor that seemed to influence 13-year-olds' performance. In five of the seven exercises, performance with a calculator was lower than without the calculator. These differences ranged from 6 to 23 percentage points. On the fifth exercise, performance remained the same. Performance on all but one of these exercises was poor, rarely reaching as high as 50 percent correct. In every case the no-response rate was much higher when calculators were available than when they were not. In most cases, the higher no-response rate accounted for the lower performance with calculators. One must wonder what caused students to fail to respond more often when a calculator was available. Is it that the decimals confused students? Were the students unable to enter or interpret the keystroking and display when decimals were involved?

Four of these same single-step exercises were also given to 17-year-olds. Performance for this age group on these exercises was fairly consistent with and without a calculator. Table 9.8 illustrates many of these findings with a sample exercise. Performance on this exercise was the highest of any of the problems in this category. Of special interest is the fact that both 13- and 17-year-olds scored higher on this exercise without a calculator than they did when using a calculator. This may reflect earlier documented difficulties students had in working with decimal numerals on the calculator.

Table 9.8

Results on a One-Step Word Problem

Exercise	Percent Responding					
	Age 9		Age 13		Age 17	
	NC*	C	NC	C	NC	C

MENU			
Hamburger	.85	Milk	.20
Hot Dog	.70	Soft Drink	.15
Grilled Cheese Sandwich	.55	Milk Shake	.45
French Fries	.40	Ice Cream	.40

Sue had a hot dog, french fries, and milk.
How much did she spend?

	Age 9 NC	Age 9 C	Age 13 NC	Age 13 C	Age 17 NC	Age 17 C
○ $1.20	9	4	2	2	1	2
● $1.30	57	62	92	77	95	87
○ $1.40	9	3	2	2	2	2
○ $1.50	16	8	3	4	1	2
○ I don't know	6	10	0	2	0	1
○ No response	3	13	1	13	1	6

*NC = No calculator available; C = calculator

A wide range of multistep exercises was included in this portion of the assessment, most of which were consumer related and two of which were given to 9-year-olds. These exercises were difficult for the 9-year-olds, as seen by less than 38 percent correct on each in either format. Performance for 13-year-olds on multistep problems was consistent with performance on one-step exercises discussed earlier in that it was generally low, with the paper/pencil format showing a slight edge. Several of these exercises were not typical of everyday experience for most 13-year-olds. Students were asked, among other things, to use unit pricing to figure savings on consumer items and to compute average gas mileage.

Seventeen-year-olds were given several of these same multistep exercises. Although their performance was slightly better than that of the 13-year-olds, the calculator seems to have made little difference in most cases. On the two released exercises given to 17-year-olds and shown in Table 9.9, about one-third of them were successful on both exercises when given calculators, whereas one-fifth and one-third gave correct answers without using a calculator. About four times as many students in the calculator group as in the no-calculator group made no response. Both of these exercises required a minimum of three steps; furthermore, they included other complicating features, such as decimals and percents. Undoubtedly, all of these factors contributed to the low performance, and in one instance the calculator proved to be of no help to the students. On exercise B, however, a slight improvement in performance was observed. It is difficult to speculate why the calculator helped improve performance in this exercise, because at least three operations were needed to arrive at an answer, just as in exercise A. However, division was not involved, and perhaps this is an explanation for the improved performance.

Table 9.9

17-Year-Olds' Performance on Multistep Problems

with and without a Calculator

| | Percent Correct | |
Exercise	NC*	C
A. Jerry bought an old Ford for $900. He paid $200 down and borrowed the rest. The total finance charge was 10% of the loan. He paid off the loan and finance charge in 10 equal installments. How much was each installment?	33	32
B. [Newspaper ad describes special] Regular price: Parts $32.50 Labor $42.50 Special: 15% off parts and labor		
What would be the total cost of the Tune-up Special?	21	36

*NC = No calculator available; C = calculator

Many people have suggested that with a calculator a student is free to explore possible routes to the solution of word problems. They have conjectured that perhaps this freedom from the often laborious task of computing would encourage students to develop and refine their problem-solving techniques. On these multistep exercises, however, perhaps because of the time constraint, the calculator proved to be of little help.

Several exercises required more than direct translation of the displayed result. For example, in an open-ended exercise similar to the following, 9-year-olds were not able to deal with decimals:

A class of 26 fourth-grade students is going to visit a museum. They will be driven there in cars. Each car can hold 4 students. How many cars are needed?

An examination of this exercise shows it is not a simple problem on the calculator but, in fact, requires thoughtful interpretation of the result. Three percent of the 9-year-olds with a calculator answered correctly. Twelve percent reported the display (6.5) as the number of cars needed, and 7 percent simply reported 65 as the number of cars needed, clearly an unreasonable answer. This latter group simply ignored the decimal, which is consistent with what many in this age group did on similar computation exercises involving decimals.

Thirteen- and 17-year-olds also had difficulty with similar exercises, as illustrated in Table 9.10. The poor performance on this exercise—both with an and without a calculator—reflects the failure to carefully think through this problem. Only a whole number of baseballs would make an appropriate answer and no more than 23 should be left over, but only 6 percent of the 13-year-olds and 19 percent of the 17-year-olds with a calculator answered correctly. Furthermore, many students in both age groups simply reported the quotient without any regard for the reasonableness of this answer.

Table 9.10

Performance on a Division Exercise with Decimal Remainder

Exercise		Percent Correct Age 13	Percent Correct Age 17
A man has 1310 baseballs to pack in boxes which hold 24 baseballs each. How many baseballs will be left over after the man has filled as many boxes as he can?	No calculator	29	--
	Calculator	6	19

This problem required students to think carefully about their answers and not mechanically record the result shown on the calculator display. In fact, this problem is more difficult to solve with a calculator, since it requires several distinct steps after the quotient is displayed. This exercise again illustrates the need for students to understand clearly what the quotient is and what it means. In particular, when a quotient involves a decimal, careful instruction should address how this value is to be interpreted. What does the decimal mean? How is it related to the remainder?

Summary

The hand calculator portion of this mathematics assessment attempted to answer the following question:

How well can students solve different types of mathematics problems when familiarized with a simple, four-function calculator?

Answering this question with any degree of certainty is hazardous, since it is uncertain whether students were familiar with the calculator they used. Would performance have been appreciably different if the initial instruction had given more thorough coverage to special input and output resulting from calculator computation? This instruction might have been extended to include discussion of decimal displays, instructions for computing with integers, special keystroking sequences, or discussion of remainders as they appear on a calculator. Indeed, this instruction would have required a greater amount of time, but it might have ensured that students were more familiar with the calculator.

Even with the limited amount of instruction provided, the data suggest that students do perform routine computation better with the aid of a calculator. This is encouraging. For if the calculator can be made available to students as an efficient, accurate computational tool, then less attention can be given to paper/pencil computation skills and more of the mathematics time can be spent on other important areas of the curriculum.

Clearly, one of the most significant results from the calculator assessment involves problem solving. Comparisons of performance of all ages on problem solving with and without the aid of a calculator provide very convincing evidence that calculators do not solve problems, people do. Strategies that take special advantage of the calculator, such as trial and error, must be introduced and developed. Problem solving requires far more than computation. It demands understanding, the correct choice of operations, and the selection of values to operate within a particular order. It is then and only then that calculators become a useful tool.

Computers

Computers have become an integral part of our society, and there has been a corresponding emphasis on computer literacy as an important objective of mathematics instruction. In anticipation of future emphasis on computer literacy and in recognition of this as a topic on which performance should be monitored, the National Assessment of Educational Progress included exercises that dealt with several aspects of computer literacy in the 1977-78 mathematics assessment. The exercises included both cognitive and affective exercises and were administered only to 13- and 17-year-olds.

Overview of Results

Following are the major observations from the data:
1. Students had a great deal of difficulty reading and interpreting flowcharts. Performance varied inversely with the complexity of the chart.
2. Both 13- and 17-year-olds were unsuccessful at reading and interpreting a simple BASIC computer program. Levels of performance were below what would be expected by chance alone.
3. Less than 15 percent of either age group had had the opportunity to study computer programming either in or out of their mathematics classes, although around two-thirds of both groups thought a knowledge of computer programming would be useful.
4. Students appeared to have some knowledge of the kinds of simple tasks computers are able to perform but seemed to lack awareness of more sophisticated computer applications.
5. Results showed that students were undecided in their judgments of the possible impact of computers on society, although over three-fourths of both groups felt that computers were ultimately going to be responsible for the operation of "most things."

Cognitive Exercises

Four cognitive exercises assessed students' knowledge of two aspects of computer literacy, flowcharting and computer programming. The two flowcharting exercises were similar in that each asked students to determine the output of a given flowchart. Table 9.11 presents results for one of these exercises. The steps in this flowchart required that students understand an assignment statement (T = T + 3) and a loop process that required a decision about whether an obtained number was less than 35. The low levels of performance (about what would be expected by chance alone) and the high percentages of students who responded "I don't know" to these exercises suggest that students have either had little opportunity to develop knowledge about flowcharts in general or, more specifically, about flowcharts containing a loop.

Performance on the other flowchart exercise was even lower, with 9 and 7 percent, respectively, of the 13- and 17-year-olds responding correctly, and 55 and 60 percent answering "I don't know." The flowchart in this exercise was somewhat more complicated than that of the previously described exercise and also contained a loop, which may account for some of the increased number of "I don't know" responses.

Interpreting a computer program presented in a computer language (BASIC) was the other aspect of the cognitive dimension of computer literacy that was assessed. BASIC was chosen because it is the most common computer language taught in secondary schools.

Table 9.11

Flowcharting Exercise and Results

This is a flowchart:

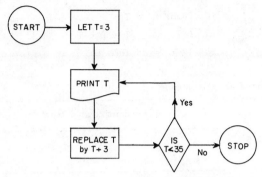

What would be printed in the output?

			Percent Responding			
Age 13	5	11	10	7	11	9	47
Age 17	6	13	7	6	22	6	39

○	○	○	○	●	○	○
3	6	33	3	3	3	I don't
			3	6	6	know
			3	9	9	
			3	12	12	
			3	15	15	
			3	18	18	
			3	21	21	
			3	24	24	
			3	27	27	
			3	30	30	
			3	33	33	
					36	
					39	
					42	
					:	

Two programming exercises were included; the results obtained on the released exercise are presented in Table 9.12. Note that this programming exercise and the flowchart exercise presented in Table 9.11 were designed to be
parallel exercises (as were the two unreleased exercises). There was little
variation in performance on these two exercises by either age group. The
levels of performance on the other BASIC exercise were comparable; 13 and 10
percent of the 13- and 17-year-olds were correct, with percentages of "I don't
know" responses totaling 57 and 64 percent, respectively.

The high percentages of "I don't know" responses on the four cognitive exercises suggest that most students recognized their lack of knowledge about the
subject of computers and chose not to even attempt the exercises. The results
are substantiated by background data and are not surprising when only 14 and 12
percent of the 13- and 17-year-olds, respectively, indicated they had studied
mathematics through computer instruction. Seventy-one and 62 percent of the
13- and 17-year-olds thought that computers would be useful in teaching mathematics, although only 12 and 25 percent had access to computer terminals for
learning mathematics in their schools.

Table 9.12

Programming Exercise and Results

This is a BASIC program.

```
 5   LET        T = 3
10   PRINT      T
15   LET        T = T + 3
20   IF T  <  35 THEN 10
25   END
```

What would be printed in the output?

	Percent Responding						
Age 13	5	3	8	8	14	9	54
Age 17	4	4	3	4	21	5	57
	○	○	○	○	●	○	○
	3	6	33	3	3	3	I don't
				3	6	6	know
				3	9	9	
				3	12	12	
				3	15	15	
				3	18	18	
				3	21	21	
				3	24	24	
				3	27	27	
				3	30	30	
				3	33	33	
						36	
						39	
						42	
						.	
						.	
						.	

The results indicate not only that most 13- and 17-year-old students have not used computers in their mathematics classes but that most have had little experience of any kind with computers. Only 8 and 13 percent of the 13- and 17-year-olds said they knew how to program a computer, although around two-thirds of both groups thought computer programming would be a useful topic to study in mathematics classes. A large majority of both age groups had neither used a computer nor written a computer program to perform such tasks as solving a linear programming problem, playing games, or analyzing statistical data. This lack of experience makes the low levels of performance on cognitive computer exercises that dealt with flowcharting and programming understandable.

Affective Exercises

Several exercises in the mathematics assessment ascertained 13- and 17-year-old students' awareness of potential uses of computers, their beliefs about the impact of computers on society, and other commonly held beliefs about computers relative to computer mystique.

Use of computers. Table 9.13 presents the results of an exercise that assessed students' opinions about the potential usefulness of computers in various situations.

Table 9.13

Results on a Computer-Awareness Exercise

| How much would a computer help in: | Age | Percent Responding Computer-- | | |
		Helps a lot	Helps a little	Does not help
A. Calculating and printing bills for a credit card company	13	72	24	3
	17	84	14	1
B. Keeping track of a car maker's inventory	13	48	42	10
	17	60	35	5
C. Telling which food tastes best	13	3	6	90
	17	1	6	93
D. Predicting election results	13	41	35	24
	17	43	42	16
E. Alphabetizing lists of names	13	66	27	7
	17	79	16	5
F. Writing a novel	13	7	18	75
	17	2	11	87
G. Diagnosing diseases	13	25	33	42
	17	24	43	33

On parts A, B, and E of this exercise, more 17- than 13-year-olds thought a computer would help "a lot" on performing the task. Since the particular tasks are representative of activities that are frequently accomplished by means of computers, the results probably reflect the 17-year-olds' knowledge of uses of computers. Both age groups were in agreement that computers would not be very useful in deciding which food tastes best or in writing a novel, and in both age groups opinion was divided with respect to the perceived usefulness of computers in the prediction of election results or the diagnosis of diseases. Thus, it appears students are aware of the routine uses of computers encountered in their everyday experiences but do not realize that computers can be used in solving more complex problems. For example, they do not appreciate the range of possibilities involving computer modeling or statistical analyses that are opened up by computers.

Beliefs about computers. A fourteen-part exercise asked the students to respond along five-point scales of "strongly disagree" to "strongly agree" to statements that reflected specific beliefs and feelings about computers. Students were knowledgeable about the simple mechanics of computer operation; over 85 percent of both age groups knew that computers had to be programmed to follow precise instructions and to store information. Around two-thirds of both groups recognized that a computer operator was not necessarily a mathematician.

Table 9.14 summarizes some other results obtained on this exercise. As the table shows, the students were fairly consistent in their responses to statements A and B, although statement A generated the greatest disparity between the two age groups on any of the statements.

More than any of the other parts of this exercise, statements C, D, and E presented in Table 9.14 represent students' value judgments about the possible impact of computers on society. Results on parts C and D of the exercise showed a remarkable agreement between the two age groups in their responses and a great deal of divided opinion within the groups. Perhaps the most surprising results were obtained on part E, however, on which over three-fourths of both age groups agreed that computers were ultimately going to be responsible for the operation of almost everything. Although most students in both age groups felt it inevitable that "someday most things will be run by computers," their responses to other statements indicated that they did not feel particularly threatened by that possibility.

Table 9.14
Responses to Statements of Feelings about Computers

Statement		Age	Percent Responding		
			Disagree	Undecided	Agree
A.	Computers are suited for do- ing repetitive, monotonous tasks.	13 17	14 11	39 64	47 25
B.	Computers have a mind of their own.	13 17	56 67	14 15	30 18
C.	Computers dehumanize society by treating everyone as a number.	13 17	31 31	37 28	32 41
D.	The more computers are used, the less privacy a person will have.	13 17	42 40	22 24	35 35
E.	Someday most things will be run by computers.	13 17	10 9	15 12	76 79

The influence of computers on the job market was reflected in other statements of the exercise. Around half of both age groups felt that "computers will probably create as many jobs as they eliminate," and slightly more than half of both groups agreed that having a knowledge of computers would help an individual get a better job. A majority of respondents also believed that computers could help make mathematics more interesting, although about one-fourth of both groups were undecided with respect to the interest-generating capability of computers.

Summary

Overall, the results on these exercises indicate that most 13- and 17-year-old students have had little firsthand experience with using or programming computers either inside or outside the mathematics classroom. Their lack of experience does not reflect, however, a lack of basic knowledge about how computers operate or a lack of opinions about computers. Whether or not the patterns of response are indicative of acceptable levels of computer literacy is difficult to ascertain. Since so few students were actually knowledgeable about computer operation, the students might have been responding on the basis of their own perceptions of computer mystique rather than from any carefully reasoned point of view. Since this may have been the case, any conclusions about levels of computer literacy among students would, at best, be tenuous.

10

Affective Responses to Mathematics

Desired outcomes of school mathematics instruction usually include both achievement-related goals and the development of positive attitudes toward mathematics. To gain information about students' affective responses to mathematics, the second NAEP mathematics assessment included a number of exercises designed to assess students' attitudes toward the subject and their perceptions of different aspects of mathematics learning.

The attitude exercises represented a different approach to attitude measurement from that employed in other large-scale investigations in that no effort was made to form item scales. Rather, items were grouped into categories that reflected their content as well as the context in which students had encountered mathematics.

Three categories of attitude exercises were given to all age groups: (1) mathematics in school, (2) mathematics and oneself, and (3) mathematics and society. A fourth category, mathematics as a discipline, was given only to 13- and 17-year-olds. All exercises administered to the 13- and 17-year-olds were identical. Those exercises given to the 9-year-olds, while reflecting the same categories, differed slightly with respect to content and format.

A word of caution is in order. As with any effort to measure attitudes, the validity of the measures is subject to some question. Only one method of gathering data, student self-report, was employed. The validity of self-report measures is subject to a number of threats; perhaps most crucial is the desire on the part of the students to present themselves in the best possible light. Since responses are directly under the students' control, they may have responded with what they considered "correct" answers rather than their true feelings. It is also questionable whether the students' interpretation of a topic or activity was the same as that intended by exercise developers; there was no means of ascertaining any differences in interpretation. Problems such as these are inherent in any effort to measure attitudes. Although such problems are potentially serious, the data are still useful for the information they provide relative to students' affective responses to mathematics.

Mathematics in School

The mathematics in school category contained items that ascertained students' feelings about mathematics as a school subject in relationship to other school subjects, particular items of mathematics content (e.g., doing addition problems), and specific classroom activities (e.g., using a mathematics textbook).

Overview of Results

Following are the major observations for which supporting data will be presented:

1. Thirteen-year-olds rated mathematics as the most important of five school subjects; mathematics and English were rated equal in importance by the 17-year-olds.

133

2. For the 9-year-olds, mathematics was the best-liked of five academic subjects; mathematics was the second best-liked subject of the 13-year-olds and the least-liked subject of the 17-year-olds.

3. The 9-year-olds rated mathematics as being about midway between easy and hard. The 13-year-olds thought mathematics was the easiest of their academic subjects, and the 17-year-olds thought it was the hardest subject.

4. Students at all age levels perceived their role in the mathematics classroom as primarily passive.

5. Most specific content topics within mathematics were rated important by all age groups; notable exceptions were the generally lower ratings of importance given to geometric topics and, for the older students, doing proofs.

6. Ratings of liking for specific topics and the ease or difficulty of the topics were very similar for the 13-year-olds; however, the 9- and 17-year-olds appeared to differentiate between their liking for a topic and its difficulty.

Discussion of Results

The school subject comparison included science, social studies, mathematics, English, and physical education at ages 13 and 17, and science, social studies, mathematics, reading, spelling, and physical education at age 9. Students were asked to indicate how much they liked or disliked each subject and how easy or how hard the subject was for them. The older respondents were also asked to rate the subjects according to how important or unimportant they thought the subjects were. Mathematics was rated important or very important by about 90 percent of the 13- and 17-year-olds. Only English, which was considered important by 90 percent of the 17-year-olds, received as high a rating.

Physical education was the best-liked subject in all age groups (liked by around 80 percent of each group). Mathematics was the favorite academic subject for the 9-year-olds (liked by 65 percent), second for the 13-year-olds (liked by 69 percent, which was fewer than the number who liked science), and last for the 17-year-olds (liked by 59 percent and disliked by 32 percent).

On the easy-hard dimension, 42 percent of the 9-year-olds thought mathematics was easy, 40 percent rated it in between, and 13 percent said it was hard. An interesting phenomenon occurred in the ratings of mathematics made by the older respondents. Among the academic subjects, mathematics was rated easy by more of the 13-year-olds and hard by more of the 17-year-olds than any other subject. In fact, mathematics was the only subject for which the pattern of responses along this dimension changed from age 13 to age 17.

Classroom activities. The respondents were presented with a list of classroom activities (e.g., taking mathematics tests) and asked to rate them according to how often they had done the activity in their mathematics courses (often, sometimes, never). These and other activities were also rated along dimensions of like-dislike and useful-useless, as shown in the sample exercise in Figure 10.1. The illustrated exercise was given to 13- and 17-year-olds; 9-year-olds responded along three-point scales instead of five-point scales.

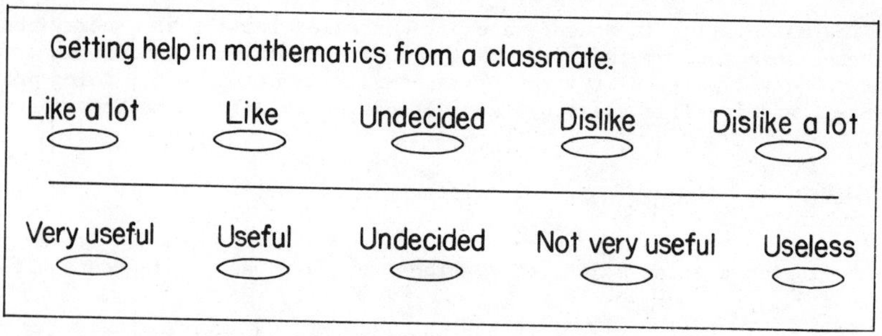

Getting help in mathematics from a classmate.

Like a lot   Like   Undecided   Dislike   Dislike a lot

Very useful   Useful   Undecided   Not very useful   Useless

Figure 10.1

The item format resulted in a fairly large percentage of students giving no response along the useful dimension, particularly for the 13-year-olds (averaging around 30 percent). While the percentage of students who did not respond is so large that definite conclusions about results cannot be drawn, results are indicative of trends and will therefore be reported.

The activities can be classified as student-centered, classmate-centered, teacher-centered, and "other." Table 10.1 summarizes students' perceptions of

Table 10.1

Indicated Frequency of Selected Classroom Activities

	Activity	Age	Percent Responding Often	Sometimes	Never
STUDENT-CENTERED	Taken mathematics tests	9	44	46	9
		13	61	37	1
		17	63	33	3
	Done mathematics homework	9	43	45	12
		13	67	29	3
		17	57	36	6
	Worked mathematics problems alone	9	71	22	7
		13	81	17	1
		17	80	19	1
	Worked mathematics problems at the board	9	39	54	7
		13	33	58	9
		17	27	60	13
	Used your mathematics textbook	9	75	18	6
		13	81	14	5
		17	87	11	3
	Done worksheets	9	71	27	2
CLASSMATE-CENTERED	Helped a classmate do mathematics	9	14	56	29
		13	11	71	17
		17	15	72	13
	Gotten help from a classmate on mathematics	9	9	61	30
		13	8	74	17
		17	17	74	8
	Discussed mathematics in class	9	49	38	12
		13	58	38	4
		17	50	43	7
	Worked mathematics problems in small groups	9	16	49	35
		13	9	47	44
		17	12	59	28
TEACHER-CENTERED	Listened to the teacher explain a mathematics lesson	9	85	11	3
		13	81	16	2
		17	78	19	2
	Watched the teacher work mathematics problems on the board	9	78	18	4
		13	76	21	2
		17	79	18	3
	Gotten individual help from the teacher on your mathematics	9	21	67	11
		13	17	71	10
		17	18	70	11

the frequencies of the separate activities in the first three of these categories. As the table shows, "using a mathematics textbook" was perceived to be the most frequently occurring activity among the student-centered activities, with "working mathematics problems alone" a close second. On the classmate- and teacher-centered activities, indicated frequencies were consistent across age groups except for the activity of "getting help from a classmate in mathematics," which apparently occurs more often in the mathematics classrooms of older students. Notice that students perceive that listening to and watching the teacher do mathematics are by far the most frequently occurring activities within their mathematics classrooms.

A majority of all age groups indicated a liking for working mathematics problems at the board, working problems alone, and using a textbook. Doing mathematics homework was the most disliked activity. Twenty, 38, and 49 percent of the 9-, 13-, and 17-year-olds, respectively, indicated they disliked doing homework. It may be that the students, particularly the younger ones, were somewhat hesitant to express dislike for any of the activities. For example, 46, 49, and 35 percent of the 9-, 13-, and 17-year-olds, respectively, said they liked taking mathematics tests, with 8, 21, and 30 percent, respectively, expressing dislike. All of the student-centered activities were rated useful by a majority of respondents.

Results on the like-dislike and useful-useless dimensions of the teacher-centered activities showed that a slight majority of all age groups liked the teacher-centered activities and found them useful. The classmate-centered activities were generally rated lower on both dimensions than the teacher-centered activities.

The "other" classroom activities included such things as choosing the mathematics to be studied and using manipulative materials in the mathematics classroom. Results suggest that students at all ages feel they have had little opportunity for input concerning the choice of material to be covered and have spent little time in laboratory types of activities. Sixty-five, 76, and 80 percent of the 9-, 13-, and 17-year-olds said they had never participated in such activities.

The results obtained on this portion of the exercises suggest that students perceive their role in the mathematics classroom to be primarily passive. They feel that they spend a great deal of time watching the teacher work problems and a lot of time working problems from the textbook on an individual basis. Outside of class discussions, students perceive very little interaction among class members relative to the mathematics being studied.

Content topics. The 9-year-olds were presented with ten content topics and were asked to rate them along three-point dimensions of easy-hard, like-do not like, and important-not important. Item format was similar to that illustrated in the previous section. Five of the topics dealt with computation, one with mathematics word problems, and four with geometry and measurement.

Nine of the ten topics were rated important by at least 70 percent of the respondents; "learning about circles, triangles, and other shapes" was seen as important by only 52 percent. Table 10.2 summarizes the ratings made by the 9-year-olds along the liking and difficulty dimensions.

As Table 10.2 shows, learning about circles, triangles, and other shapes received the highest percentage of easy ratings. This result is in contrast to the fact that the topic was also rated lowest in importance among the ten topics. Weighing objects with a scale received the highest percentage of "hard" ratings of the geometry and measurement activities for the 9-year-olds.

Among the computation activities (doing addition problems, doing subtraction problems with borrowing, checking a subtraction problem by adding, learning multiplication or times tables, and dividing one number by another), adding and checking subtraction were rated easy by around two-thirds of the 9-year-olds. Learning multiplication tables was the topic most often selected as hard (23 percent), although solving mathematics word problems received the lowest percentage of easy ratings (31 percent).

Table 10.2

Ratings by 9-Year-Olds on Content Topics

Activity	Percent Responding			
	Like	Dislike	Easy	Hard
Doing addition problems	65	10	65	9
Doing subtraction problems (w/ example)	47	20	59	12
Checking subtraction by addition	51	16	66	9
Learning multiplication or times tables	56	14	39	23
Dividing one number by another	50	18	44	16
Solving mathematics word problems	45	18	31	18
Learning about money	68	7	52	9
Learning how to measure things with a ruler	59	11	61	9
Learning how to weigh objects in school with a scale	58	15	39	11
Learning about circles, triangles, and other shapes	63	11	82	4

Learning about money appeared to be the best-liked of the ten topics, although doing addition problems was a close second. Solving word problems was apparently the least-liked topic, although percentages on this and the topic of doing subtraction problems with borrowing were very close.

The 13- and 17-year-olds were presented with seventeen content topics and were asked to rate them along five-point scales of importance, difficulty level, and liking. The topics can be divided into categories of computation, geometry and measurement, estimation, and "others." (For ease in reporting results, the two extreme response options on each end of each scale have been combined into one.)

Table 10.3 summarizes the percentages of importance, difficulty, and liking ratings made by the 13- and 17-year-olds on all content topics. As the table shows, ratings on the importance dimension were generally high for all topics except for the topic of doing proofs. Working with decimals, checking answers to problems, estimating measurements, and solving equations were the topics rated highest in importance.

On the easy dimension, working with whole numbers was the topic rated highest. About one-fourth of both age groups were undecided as to whether the geometry/measurement topics and the estimation topics were easy or hard. Doing proofs and programming computers received the highest percentages of undecided votes of all topics along this dimension (an average of over one-third of each age group). Working with metric measures received the highest percentages of hard ratings in both age groups; checking answers received the highest percentages of easy ratings in both groups. The largest differences in ratings between the 13- and 17-year-olds occurred on the topic of solving word problems, with over twice as many of the older respondents rating the topic hard.

The topic of solving word problems created the largest differences between the older groups in ratings along the liking dimension as well. Although around one-fourth of each group said they were undecided as to whether they liked or disliked solving word problems, about twice as many 13-year-olds as 17-year-olds liked doing so, with the proportion reversed on the dislike ratings. It may be that part of this increase in dislike for word problems from age 13 to age 17 is due to experiences with solving word problems in algebra, an area notorious among students as being difficult. Evidence from the easy ratings lends some support to this assertion.

Table 10.3

Ratings on Content Topics

	Topic	Age	Percent Responding Important	Easy	Like
COMPUTATION	Working with whole numbers	13	78	75	64
		17	83	80	66
	Doing long division	13	70	50	40
		17	69	63	40
	Working with fractions	13	78	52	46
		17	81	58	46
	Working with decimals	13	86	62	59
		17	85	71	50
	Working with percentages	13	77	45	42
		17	80	47	34
GEOMETRY/ MEASUREMENT	Learning about geometric figures	13	63	32	33
		17	56	35	29
	Measuring lengths, weights, or volumes	13	75	45	40
		17	83	50	40
	Working with metric measures	13	73	34	36
		17	73	31	28
ESTIMATION	Estimating answers to problems	13	63	50	44
		17	66	48	37
	Estimating measurements (lengths, weights, areas, etc.)	13	80	50	43
		17	85	49	38
OTHER	Solving word problems	13	74	57	59
		17	69	37	32
	Doing proofs	13	48	29	26
		17	43	26	19
	Solving equations	13	82	59	56
		17	80	54	46
	Memorizing rules and formulas	13	69	34	31
		17	57	28	20
	Using charts and graphs	13	74	61	55
		17	71	63	52
	Checking the answer to a problem by going back over it	13	81	68	40
		17	89	76	43
	Programming a computer	13	63	19	30
		17	60	15	25

In general, it appears that more 17-year-olds than 13-year-olds disliked the topics. These results reflect the general decline from age 13 to age 17 in positive feelings toward mathematics reported earlier.

The results obtained on the content topic evaluations suggest several general observations. First, a majority of the students perceived most of the content topics as important. Comparatively speaking, the relatively low rating of importance assigned to the study of geometric ideas by the 9-year-olds is of some concern. The study of geometric ideas provides a natural introduction to basic ideas of mathematical applications and should be an essential feature of the mathematics curriculum. It may be that students are not being made aware that their study of geometric ideas is a study of the world around them.

Second, the 9- and 17-year-olds' ratings of topics as easy or hard was related to their performance on those topics, whereas the 13-year-olds did not appear to make such a distinction. For example, performance on solving multi-step and nonroutine problems was generally low; the 9- and 17-year-olds rated this topic difficult, but 57 percent of the 13-year-olds thought the topic was easy. Along this same line, the 9- and 17-year-olds appeared to make more distinctions than the 13-year-olds between ratings of liking and difficulty. For the 13-year-olds, liking and ease of topics corresponded fairly closely, but this was not necessarily true for the 9- and 17-year-olds. Perhaps the most striking observation to be drawn from these data is that students can, and do, evaluate specific topics within mathematics along several dimensions (e.g., importance, difficulty, and liking). Further, their evaluations may or may not correspond to their teachers' a priori expectations.

Summary

Results presented in this section have suggested that students at all age levels perceive mathematics to be an important subject to study, both globally and in terms of specific topics. Younger students like mathematics as a subject, but liking declines as age increases. Further, data from the frequency of classroom activities exercises indicate that students view their role in the mathematics classroom to be much more passive than active. Other results suggested, however, that the students do not particularly dislike taking a passive part in their mathematics classes.

The remainder of this discussion will focus on results of the categories of mathematics and oneself, mathematics and society, and mathematics as a discipline. All of the items in these categories were statements to which students responded along Likert-type scales of disagree to agree that included an "undecided" response option. The 9-year-olds responded along three-point scales; the 13- and 17-year-olds had five-point response scales. For ease in reporting results for the 13- and 17-year-olds, the two extreme options on either end of the scale have been combined into one; for example, "strongly agree" and "agree" will be reported as "agree."

Mathematics and Oneself

Overview of Results

Among the exercises of the mathematics and oneself category were statements that reflected students' desires to succeed in mathematics and their self-concept in mathematics. Following are the general results:

1. A large majority of students at all three age levels expressed feelings of wanting to be successful in their study of mathematics.
2. Over half of all three age groups thought they were good at mathematics and said they enjoyed the subject.
3. There was a slight decline in overall favorableness toward mathematics from age 13 to age 17.

Discussion of Results

Three statements assessed students' desires to be successful in their study of mathematics: (1) I really want to do well in mathematics; (2) My parents really want me to do well in mathematics; and (3) I am willing to work hard to do well in mathematics. Three-fourths of the 9-year-olds and over nine-tenths of the 13- and 17-year-olds said they wanted to do well in mathematics and were willing to work hard enough to achieve their goal. Ninety-five percent of the two younger groups and 87 percent of the 17-year-olds perceived that their parents wanted them to do well in mathematics. Around 90 percent of the 13- and 17-year-olds also agreed that a good grade in mathematics was important to them.

Other results showed that a majority of all three groups felt that they were fairly good mathematics students. Fifty-five percent of the 9-year-olds felt that they were good at working with numbers, and an additional 40 percent thought they were sometimes good at the task. Sixty-five and 54 percent of the 13- and 17-year-olds, respectively, felt they were good at mathematics, although 9 and 22 percent felt they were not. Thirty-nine percent of the 9-year-olds said they always understood what was being discussed in their mathematics classes, and an additional 57 percent said they sometimes understood. A majority of students in both of the older groups felt that they usually understood what was going on in their mathematics classes.

About half of the 9- and 17-year-olds and two-thirds of the 13-year-olds claimed they enjoyed mathematics. These relatively high percentages may surprise some mathematics teachers, but they are consistent with results obtained on the school subject comparison.

Slightly over one-fourth of both older groups said they were taking mathematics because they had to, and this same proportion was undecided as to whether they would like to take more mathematics. Although 63 percent of the 17-year-olds did not feel that they were taking mathematics because they had to, only 38 percent said they wanted to take more mathematics. The 13-year-olds were more consistent in their responses to these statements than the 17-year-olds. These results, although certainly of interest to mathematics teachers, are somewhat difficult to evaluate, since comparable information was not obtained about other school subjects.

Summary

Across all age groups, the results indicate that students perceive themselves as motivated to succeed in their study of mathematics, as being concerned about making good grades in the subject, and as usually understanding the material presented to them in a mathematics class. They claimed to enjoy the subject, although the older students expressed some doubt about how much mathematics beyond requirements they would pursue. Some teachers may find these results a little surprising when weighed against their experiences with their own students. Any discrepancies between teachers' perceptions and the students' expressed perceptions could be due to a number of reasons, not the least of which might be the tendency of students to give "right" answers, probably a strong tendency with the statements in this category. In spite of the assurances of anonymity given the students, they may have been hesitant to admit that they did not enjoy mathematics or that they had trouble following what was going on in class.

Mathematics and Society

Overview of Results

The exercises in the mathematics and society category dealt with students' perceptions of the usefulness of mathematics to themselves as individuals and to broader societal concerns. The exercises focused on the usefulness of mathematics and the job-related importance of mathematics. The general observations of the results include the following:

1. Students at all age levels were consistent in their belief that mathematics is a practical subject that is useful in helping solve everyday problems.

2. A large majority of students at all age levels felt that a knowledge
    of mathematics was important in order to get a "good" job.

Discussion of Results

Over three-fourths of the 13- and 17-year-olds and two-thirds of the 9-
year-olds felt that mathematics was useful in helping solve everyday problems.
Further, around 80 percent of the older respondents thought that most mathematics had some practical use. Students were consistent in their belief that
mathematics was useful; a large majority of all age groups felt that they could
not get along very well in everyday life without using some mathematics.

Most students felt that some knowledge of mathematics was important if a
person was to get a "good" job, but the 9-year-olds were unable to determine if
people used mathematics in their jobs. In response to the statement "Most people do not use mathematics in their jobs," 39 percent of the 9-year-olds disagreed, 35 percent agreed, and 26 percent were undecided. There was also a
lack of consensus about whether they themselves would "like to work at a job
that lets me use mathematics." Forty-four percent agreed that they would like
to do so, 22 percent disagreed, but 34 percent were undecided. (These statements were not given to the older respondents.) Over 80 percent of all three
age groups thought a knowledge of arithmetic was important for the job seeker,
but only 72 percent of the 13-year-olds and 46 percent of the 17-year-olds
thought knowledge of algebra and geometry important. It may be that experience
with algebra and geometry has not convinced the 17-year-olds that the subjects
are useful, while the 13-year-olds, who have not taken these courses, believe
these subjects are useful.

Mathematics as a Discipline

Overview of Results

With the exception of two exercises dealing with sex-related stereotyping
of mathematics that were also given to the 9-year-olds, all exercises in the
mathematics as a discipline category were given only to 13- and 17-year-olds.
These exercises dealt with mathematics as a fixed or changing subject, mathematics as a process-oriented versus rule-oriented subject, and other general
perceptions of the subject and the people who study it. The following general
observations are based on the data from this portion of the assessment:

1. Students at all age levels do not perceive mathematics as either
    male- or female-dominated; that is, mathematics is a subject for
    everyone.
2. A large majority of the older students felt that mathematics was
    rule based but also that knowledge of process was as important as
    getting the answer.
3. The 13- and 17-year-olds appear to have little perception of what
    mathematicians do.
4. Seventeen-year-olds appear to recognize that mathematics is a devel-
    oping subject, whereas many 13-year-olds do not.

Discussion of Results

It is not uncommon to find references to the notion that "mathematics is
a subject for males" used to help explain why females often choose not to pursue the study of mathematics beyond requirements. In an effort to gain information about students' perceptions of this notion, two statements were included
in this portion of the assessment that dealt with this purported sex-related
stereotyping of mathematics. Results are presented in Table 10.4. As the table
shows, all groups of students disagreed with both statements, with the older

students disagreeing even more strongly than the 9-year-olds. These results suggest that students do not perceive mathematics to be a subject more suitable for one sex than for the other. In both instances, more females than males disagreed with the given statement; for those students who agreed, the tendency was to agree in favor of their own sex. This last result, while not surprising, should not outweigh the significant fact that the large majority of students did not perceive mathematics to be more suitable for males than females or vice versa.

Table 10.4

Students' Perceptions of the Sex-related
Stereotyping of Mathematics

| Statement | Age | Percent Responding | | | |
| | | Agree | | Disagree | |
		Males	Females	Males	Females
Mathematics is more for girls than for boys.	9	11	16	68	63
	13	2	9	80	90
	17	4	1	85	92
Mathematics is more for boys than for girls.	9	21	10	65	66
	13	3	2	89	96
	17	3	2	87	94

Other exercises in this category dealt with the older students' perceptions of mathematics as either rule oriented or process oriented; results are summarized in Table 10.5. Although the students appeared to hold divided opinions about whether learning mathematics is mostly memorizing, they almost unanimously felt that mathematics is very much rule based. Just as strongly, however, they felt that knowledge of process is as important as answer-getting. These views seem almost contradictory, but both age groups made remarkably similar responses to the statements. It may also be that students view knowing the process as equivalent to knowing the rule. One statement that did generate some disagreement between age groups was "Trial and error can often be used to solve a mathematics problem." If students really believe that doing mathematics requires following rules, then it seems that more students would disagree that trial and error could be used to solve problems.

It is interesting to speculate about students' reasons for responding to these particular questions as they did; do they really feel what they said, or were they trying to give answers they thought were correct? The students probably did not consider their responses contradictory in light of their experiences with mathematics. For the most part, their mathematics has been oriented toward computation types of activities for which there is always a rule to follow, and in order to be successful, one needs to practice following the rules. Students have also heard, however, that they must understand why rules work. Whether they really believe this or are merely paying lip service to a frequently espoused view is impossible to determine. Similarly, it is impossible to verify the validity of students' responses to the statement "Mathematics helps a person to think logically." This is an often-heard statement, and students may have simply been reflecting what they thought was the "right" response to the statement.

Table 10.6 presents results on the remaining exercises in this category. As results for part A show, students appeared to lack an understanding of what mathematicians do, although their responses to part B suggest they think men and women are probably equally qualified to become mathematicians. Responses to part D indicate that the 17-year-olds are more inclined to view mathematics as an integrated subject than the 13-year-olds are; whether or not this is due to their broader experience with the subject is not known. Results on part E suggest, however, that the 17-year-olds recognize to a greater

Table 10.5

Students' Perceptions of Mathematics as Rule or Process Oriented

Statement	Age	Percent Responding Disagree	Undecided	Agree
Learning mathematics is mostly memorizing.	13	33	18	48
	17	40	14	45
There is always a rule to follow in solving mathematics problems.	13	5	5	89
	17	8	4	88
Doing mathematics requires lots of practice in following rules.	13	12	11	77
	17	8	12	80
Knowing how to solve a problem is as important as getting a solution.	13	4	8	88
	17	3	4	92
Knowing why an answer is correct is as important as getting the correct answer.	13	4	7	88
	17	3	4	93
Justifying the mathematical statements a person makes is an extremely important part of mathematics.	13	4	31	65
	17	5	28	68
Trial and error can often be used to solve a mathematics problem.	13	13	31	56
	17	10	19	70
Exploring number patterns plays almost no part in mathematics.	13	64	22	13
	17	68	23	8
Mathematics helps a person to think logically.	13	6	20	73
	17	8	16	76

Table 10.6

Students' Perceptions of Mathematics and Mathematicians

Statement	Age	Percent Responding Disagree	Undecided	Agree
A. Mathematicians work with symbols rather than ideas.	13	24	44	32
	17	35	37	28
B. Fewer men than women have the logical ability to become mathematicians.	13	57	27	16
	17	73	20	7
C. Creative people usually have trouble with mathematics.	13	51	33	15
	17	56	34	10
D. Mathematics is made up of unrelated topics.	13	49	33	18
	17	59	29	12
E. New discoveries are seldom made in mathematics.	13	41	22	35
	17	53	28	18

extent than the 13-year-olds that mathematics is a changing subject and that new discoveries in mathematics are possible. This last result is in all probability due to the broader experience of 17-year-olds with the subject.

Summary

Results showed that none of the groups of students thought that mathematics was more suitable for one sex than the other, with the older students expressing even stronger feelings than the 9-year-olds. Older students also appeared to believe that mathematics was both rule oriented and process oriented, a position that may appear contradictory. Further, the 17-year-olds, more than the 13-year-olds, appeared to view mathematics as subject to new discoveries in the field.

11

Conclusions

The preceding chapters describe results from the second mathematics assessment of National Assessment. The purpose of this chapter is to present the broad conclusions that seem warranted from those results. We have identified several areas in which National Assessment results provide information about students' knowledge of mathematics that have particular importance for the mathematics curriculum and for mathematics instruction. The areas in which conclusions seem warranted include (1) the status of students' computational and noncomputational skills; (2) students' understanding of mathematical concepts and processes; (3) students' problem-solving skills; (4) the continued development of students' mathematical skills; (5) the need for the development of alternative computation algorithms; (6) the need to increase and extend students' enrollment in mathematics courses; and (7) the need to increase students' active involvement in mathematics classroom activities.

Caution must be observed in interpreting the results from the NAEP mathematics assessment. The assessment is not designed to identify causes of student performance, and so in order to provide some general conclusions, we have frequently extrapolated beyond the data. Other authors would possibly reach different conclusions. However, the conclusions that we have presented are generally supported by a wide range of exercises in addition to those illustrative exercises we have reported in earlier chapters.

Computational versus Noncomputational Skills

If students' performance on the second mathematics assessment is a measure of instructional emphasis, then one must conclude that the focus of most mathematics programs is on the development of routine computational skills. Students demonstrated a high level of mastery of computational skills, especially those involving whole numbers. The majority of students at all age levels demonstrated severe deficiencies in other basic skill areas, such as geometry, measurement, and probability and statistics.

Almost all students demonstrated mastery of basic number facts. About two-thirds of the 9-year-olds could perform simple addition and subtraction computations using algorithms for regrouping, and by age 13 almost all students could perform simple computations involving addition, subtraction, and multiplication. Most of the older students were successful with more difficult calculations. Students encountered greater difficulty with whole number division and operations with fractions and decimals. Performance was significantly lower, however, on exercises assessing basic noncomputational skills.

In general, the only noncomputational skills for which students demonstrated a high level of mastery were those involving simple intuitive concepts or those concepts or skills they were likely to have encountered and practiced outside of school. This was reflected in students' knowledge of basic geometry and measurement concepts, as described in chapters 6 and 7.

Understanding Mathematical Concepts and Processes

As results in the previous section showed, students are failing to master a broad range of basic skills. Further, many of the skills that they

have learned have been learned by rote, at a superficial level. <u>Students'</u>
<u>performance showed a lack of understanding of basic concepts and processes</u>
<u>in many content areas, such as measurement and computation with fractions.</u>
Almost all students could make simple linear measurements. Over 80
percent of the 9-year-olds and 90 percent of the 13-year-olds could measure
the length of a segment to the nearest inch, but only about 20 percent of the
9-year-olds and about 60 percent of the 13-year-olds could correctly determine
the measure of a line segment when it was aligned so that the left endpoint
was not at zero on the ruler. Thus, although most students would line up the
end of the segment when they were measuring it, changing the problem context
demonstrated that they did not understand the consequences of not doing so.
Students' superficial understanding was also apparent in many computation exercises. For example, students were relatively successful in multiplying two common fractions, perhaps because multiplying numerators and denominators seems to be the natural way to approach the problem. Given a simple
verbal problem that required fraction multiplication for its solution, however, fewer than one-third of the 13- and 17-year-olds gave the correct response, despite the fact that around three-fourths of both groups could correctly multiply the fractions involved. These results indicate that students
have no clear conception of the meaning of fraction multiplication and therefore could not apply their skills to solve a simple problem.
The importance of understanding may, in part, account for the difference
in the level of performance between whole number operations and operations
involving fractions and decimals. Most assessment exercises indicated that
students have learned the basic concepts underlying whole number computation
and have some notion of the place-value concepts involved in the computation
algorithms. As a consequence, performance on whole number computation exercises was generally good. The results also suggest, however, that most students do not have a clear understanding of fraction operations and appear to
operate at a mechanical level. This lack of understanding resulted in relatively poor performance on some fraction computation and is further highlighted by the serious difficulties encountered in solving simple problems
involving fraction operations.
Although the results suggest that students have at best a superficial
understanding of many mathematical concepts and processes, around 90 percent
of the 13- and 17-year-olds felt that developing understanding was an integral
part of mathematics learning, as evidenced by their agreement with the statement "Knowing why an answer is correct is as important as getting the correct
answer." Their responses may reflect their actual beliefs or may, in fact,
reflect their belief that the statement is one they have heard from their
mathematics teachers and that agreement with the statement was the "right"
response. The second alternative gains credence when one considers that
around 90 percent of both older groups agreed that "there is always a rule to
follow in solving mathematics problems." The students may be concentrating
on mastering rules to the extent of ignoring concomitant understanding, because their experience dictates that right answers, usually obtained through
rules, are rewarded.

Problem-solving Skills

One of the consequences of students' learning mathematical skills by
rote is that they cannot apply the skills they have learned to solve problems.
In general, NAEP results showed that the majority of students at all age
levels had difficulty with any nonroutine problem that required some analysis
or thinking. <u>It appears that students have not learned basic problem-solving</u>
<u>skills.</u>
A word of caution is in order in describing the performance of students
on problem solving. Problem solving is often equated with solving verbal
textbook problems, but this was not the type of problem that caused difficulty. In fact, students generally were successful in solving routine one-
step verbal problems, such as those found in typical textbooks. Results suggested that if students understood the operation involved in routine one-step

verbal problems, finding the solution presented no difficulty. They encountered difficulty only when they relied on a mechanical knowledge of a particular algorithm.

Although students could successfully identify which operation should be
used to solve most simple one-step problems, they had a great deal of difficulty analyzing nonroutine or multistep problems. In fact, given a problem
that required several steps or contained extraneous information, students frequently attempted to apply a single operation to the numbers given in the
problem.

Even when students could identify the appropriate operation to use to
solve a problem, they frequently had difficulty relating the results of their
calculation to the given problem in nonroutine situations. For example,
recall the baseball problem administered to 13-year-olds:

A man has 1310 baseballs to pack in boxes which hold 24 baseballs
each. How many baseballs will be left over after he has filled
as many boxes as he can?

Twenty-nine percent recognized that the remainder (14) of the division calculation was the correct response, but 22 percent gave the quotient (54) as
their answer. This error occurred because the problem required students to
do more than routinely identify an appropriate operation and perform the calculation. Apparently problem solving involves only these two steps for too
many students.

Students had difficulty with problems in many instances because they
had not developed good strategies for solving those problems. For example,
when faced with problems that contained extraneous data, students often attempted to incorporate all of the numbers given in the problem into finding
their solution. Other results showed that students did not draw pictures to
help themselves understand problems, nor were they able to apply their knowledge of related problems to solve a given problem.

The assessment results indicate that the primary area of concern should
not be with simple one-step verbal problems, but with nonroutine problems that
require more than a simple application of a single arithmetic operation. Part
of the cause of students' difficulty with nonroutine problems may lie in our
everemphasis on one-step problems that can be solved by simply adding, subtracting, multiplying, or dividing. Many articles that deal with solving verbal problems discuss problem solving in terms of choosing the correct operation. The assessment results indicate that students have relatively little
difficulty solving problems that only require them to choose the correct operation. In fact, their difficulties with nonroutine problems seem to result
from their interpretation that problem solving simply involves choosing the
appropriate arithmetic operation and applying it to the numbers given in the
problem.

Instruction that reinforces this simplistic approach to problem solving
may contribute to students' difficulty in solving unfamiliar problems. Although it can be argued that children must learn to solve simple one-step
problems before they can have any hope of solving more complex problems, an
overemphasis on one-step problems may only teach children how to routinely
solve this type of problem. It may also teach them that they do not have to
think about problems or analyze them in any detail.

Techniques designed to give children success with simple one-step problems that do not generalize to more complex problems may be counterproductive.
For example, focusing on key words that are generally associated with a given
operation provides a crutch on which children may come to rely. Such an approach provides no foundation for developing skills for solving unfamiliar
problems. Simple one-step problems may provide a basis for developing
problem-solving skills, but only if they are approached as true problem-
solving situations in which students are asked to think about the problem
and develop a plan for solving it based on the data given in the problem and
the unknown they are asked to find.

Students need to learn how to analyze problem situations through instruction that encourages them to think about problems and helps them to develop good problem-solving strategies. There is no magic formula for making

students into good problem solvers. It is clear that they need ample opportunity to engage in problem-solving activity. If problem solving is regarded as secondary to learning certain basic computational skills, students are going to be poor problem solvers.

The Continued Development of Mathematical Skills

Although problem solving and other content areas clearly require an increased emphasis in the curriculum, we do not deny the importance of computational skills. A reasonable level of computational skill is required for problem solving. We are suggesting, however, that problem solving not be deferred until computational skills are mastered. Problem solving and the learning of more advanced skills reinforce the learning of computational skills and provide meaning for their application.

It is important to recognize that <u>most computational skills are learned over an extended period of time</u>. Assessment results suggest that most skills are mastered after their period of primary emphasis in the curriculum. For example, even though a goal of most mathematics programs is that students learn subtraction facts by age 9, there was significant improvement in performance on subtraction exercises from age 9 to 13, and there even was some improvement between 13 and 17. Results also suggest that some of students' basic misconceptions may disappear as students progress through school.

These results seem to have profound implications for minimum competency programs. They suggest that rigid minimum competency programs that hold children back until they have demonstrated mastery of a given set of skills may, in fact, be depriving them of the very experiences that would lead to mastery of the particular skills.

We cannot be complacent, however, and assume that skills will naturally develop as students mature. Specific provisions must be made to practice and reinforce the development of critical skills. The skills that continue to develop--for example, addition, subtraction, and multiplication of whole numbers--are skills that are used in a variety of contexts, and so students continue to have experiences with them in the curriculum.

Although some skills will continue to develop through use in other contexts, others will not. The current high school curriculum does not take into account that many basic skills are not well developed by the time students begin instruction in algebra and geometry. For example, very few 13- or 17-year-olds have mastered percent concepts or skills, but outside of general mathematics there is very little opportunity for high school students to extend or maintain their knowledge of percent.

Development of Alternative Algorithms

The widespread availability of hand calculators has profound implications for the mathematics to be taught in the schools. Over 85 percent of the 17-year-olds in the assessment indicated that they had access to a calculator. <u>This almost universal availability of calculators has implications for the appropriate level of emphasis that computation should receive and the types of algorithms we should teach</u>.

Results indicate that in spite of the extensive instruction provided on whole number division, only half of the students are reasonably proficient in division by the time they graduate from high school. With a calculator, however, over 50 percent of the 9-year-olds and over 90 percent of the 17-year-olds could do long division correctly. This raises some serious questions as to whether the time spent drilling on division is a productive use of time and effort that might otherwise be devoted to other topics. Certainly it is clear that the current approach to teaching division is not effective for most students.

The division algorithm as well as most of the other algorithms that we teach in school are designed to produce rapid, accurate calculation procedures. Given the widespread availability of hand calculators, it would seem that the continued emphasis on developing facility with computation algorithms should not have as high a priority as it did formerly. Certainly computation is important, but what is needed are algorithms that students will remember and will be able to generalize to new situations; this brings us back to the issue of understanding. Students are more likely to remember and be able to generalize and apply algorithms if they understand how they work. Thus, it would seem appropriate to begin to shift to computational algorithms that can be more easily understood than the current ones, even if such algorithms might be less efficient.

Calculators sometimes require alternative interpretations of situations and require that students have a deeper understanding of numbers and how the operations work. They also place increased importance on estimation skills and an alertness to the reasonableness of results.

Participation in Mathematics Courses

A previous section proposed that mathematics learning is a continuous process that encompasses the entire twelve years of elementary and secondary school. Many, if not most, basic skills are not mastered by age 13 and must be reinforced and developed as part of the high school curriculum. Consequently, if we are going to improve the mathematics performance of high school graduates significantly, we must ensure that they continue to take mathematics throughout their high school program. The assessment background data indicate that this is currently not being done. The majority of 17-year-olds take only two years of high school level mathematics.

There is less difference in male and female enrollment in mathematics classes than might have been predicted from earlier studies. Since course background data were not gathered in the first assessment, it is difficult to determine whether more females are currently enrolling in mathematics courses than previously or whether earlier studies predicted a greater disparity because they surveyed a less representative sample of the population. In any event, these figures indicate that there is no significant difference in male and female enrollment in those mathematics courses that are typically taken in the first three years of high school. Significantly fewer females, however, complete the four-year mathematics sequence. This means that fewer females are prepared to take calculus in college, which seriously limits their choice of academic majors. Thus, although approximately as many females as males are taking the mathematics required to satisfy university entrance requirements, fewer females are taking the prerequisite mathematics for a major in a technical field.

There is a much greater disparity between the enrollments of blacks and whites in mathematics. The typical black takes a full year less of mathematics than the typical white. These data clearly demonstrate how inadequately we are meeting the educational needs of minority students. Improving the mathematics education of minority students should be of highest priority for the reforms of the 1980s.

Participation in Mathematics Classroom Activities

The specific types of activities in which students participate when they are in their mathematics classes is also an important factor that influences their mathematics learning. Results on certain exercises administered during the second mathematics assessment confirm the findings of other studies that students currently assume a passive role in learning mathematics.

These National Assessment results showed that students perceive their role in the mathematics classroom to be primarily passive—they are to sit and listen and watch the teacher do mathematics and then spend the rest of the time working on an individual basis on problems from the text or from worksheets. They feel they have little opportunity to interact with their classmates about the mathematics being studied. Exploratory activities and work with manipulatives are infrequent activities.

The results suggest that the current situation is one in which mathematics instruction is "show and tell" on the teacher's part, "listen and do" for the students. If active student involvement in the learning process is a desired goal of mathematics instruction, then there must be changes in approaches to teaching mathematics that will foster and encourage that involvement.

Closing Comments

The assessment results suggest that the development of routine computational skills is the dominant focus of the school mathematics curriculum and that the development of problem-solving skills is inadequate. Although we are a long way from the kind of broad mathematics program envisioned for the future, the assessment results do provide some basis for cautious optimism. It is probably fair to say that the focus of mathematics instruction has been on computation. Thus, there is evidence that students are learning what they are being taught. There is also evidence that curricular reforms can have some impact. On exercises that measured change in performance from the first assessment, there were significant gains of 10 to 20 percentage points on exercises that dealt with metric measurement. These results appear to reflect the increased emphasis on metric measurement in the curriculum over that period of time.

Improved student performance in mathematics is a goal that demands the combined efforts of teachers and administrators. The results presented have shown that there is room for much improvement, but there is hope that if we can reorganize the mathematics curriculum to address these concerns, then students' performance will improve accordingly.

Appendix A
Minorities and Mathematics

Constance Martin Anick

Thomas P. Carpenter

Carol Smith

A fundamental goal of schooling in the United States is to provide equity in education for all students. There are at least two ways that equity might be defined: (1) in terms of opportunities provided by schools, or (2) in terms of student outcomes. From the second perspective, one can argue that if any population of students is achieving significantly below its peers in the nation as a whole or is participating in fewer advanced level mathematics courses, then equity in education is not being provided. This does not imply that schools are totally responsible for differences in achievement patterns of different population groups. Many other factors within our social structure affect students' performance in school. But it is a matter of utmost concern when large segments of the population score significantly below their peers on achievement tests and fail to develop the academic skills required to enter many occupations. The elimination of these deficiencies should be a primary goal of education.

Results from the second mathematics assessment clearly document that serious inequities exist in the mathematics education of black and Hispanic students in the United States. National Assessment results are based on the performance of a carefully selected, representative national sample of over 70 000 9-, 13-, and 17-year-olds. Between 250 and 450 exercises that covered a wide range of basic mathematics objectives were administered to these students during the 1977-1978 school year. In addition to the assessment of mathematics achievement, data were gathered on a number of affective variables and on the mathematics courses taken in high school.

An important objective of National Assessment is to describe the performance of major groups within the national population. The identification of the reporting groups was based on region of the country, sex, race, level of parents' education, and size and type of community. The focus of this article is on the assessment results for blacks and Hispanics. Other minorities are not included because they were not sampled in sufficient numbers to provide reliable measures of performance. Approximately 14 percent of the sample was black, 5 percent was Hispanic, and 1 percent was identified as other minorities. These classifications were based on self-report at age 17 and on appearance and surname at ages 9 and 13.

Although the focus of this article is on blacks and Hispanics, selected results based on type of community and parental education have been included because they provide a perspective on the performance of blacks and Hispanics. Type of community is defined on the basis of occupation. Students who live in a metropolitan area where a high proportion of the adults are employed in professional or managerial positions and a low proportion are factory workers, farm workers, not regularly employed, or on welfare were classified as high metro. Students who live in areas where a high proportion of the adult population is not regularly employed or is on welfare and a low proportion is employed in professional or managerial positions were classified as low metro. The type of community classification is made for each school on the basis of the principal's estimate of the percentage of students whose parents fit in each occupational category. Clearly, this variable represents a very primitive measure of students' background which is potentially subject to a high degree of error. The level of parental education is gathered for individual students, but it is based on self-report by the students and is consequently subject to error, especially at the early age levels.

National Assessment is not designed to identify causes of particular patterns of achievement. It can only provide data on student performance and affective responses at given points in time. It does, however, give a breakdown of achievement by specific content areas and cognitive levels. Since the assessment was conducted in 1972-73 and again in 1977-78, it also provides a carefully controlled measure of change in performance.

Mathematics Achievement Results

National Assessment exercises are designed to be interpreted on an exercise-by-exercise basis, and a total score for the assessment is difficult to interpret. However, some sort of aggregation is necessary to draw any conclusions about how the mathematics achievement of minorities compares to achievement in the nation as a whole. Results for selected reporting groups are summarized in Table A.1. The numbers in the table represent the mean percent correct for all exercises administered at each age level.

Table A.1

Average Performance for Selected Reporting Categories

Category	Mean Percent Correct for All Administered Exercises		
	Age 9	Age 13	Age 17
Nation	52	54	58
Minority Group			
Black	41	39	41
Hispanic	42	43	46
Type of Community			
Low Metro	43	43	45
High Metro	60	62	68
Parental Education			
Not Graduated from High School	43	46	48
Graduated from High School	52	53	55
Post-High School	57	61	64

The results in Table A.1 indicate that both black and Hispanic performance was significantly below the national average at each age assessed. At age 9, blacks were about 11 percentage points below the national average, and the difference increased for each successive age group. At age 13 the difference was about 15 percentage points, and at age 17 it was 17 points. A somewhat similar, but less pronounced, pattern was found for Hispanics. At age 9, their average was 9 percentage points below the national average; by age 13, the difference increased to 12 percentage points, and at age 17 it was still 12 points below the national average.

Although the type of community and parental education variables may be imperfect measures of student background, the results in Table A.1 indicate that they are clearly related to student achievement. Performance for low metro students was significantly below the national average and performance for high metro was above the average. In fact, the performance of low metro students closely paralleled that of black students. Similarly, the level of parental education was directly related to student performance.

Parental education and occupational status may account for some of the disparity in minority achievement. Blacks and Hispanics were disproportionately represented in the low-scoring type of community and parental education groups and under-represented in the high-scoring groups. Thirty-three percent of the blacks and 18 percent of the Hispanics in the 17-year-old sample were in the low metro group, whereas the national average was less than 10 percent. On the other hand, 10 percent of the students in the national sample were in the high metro group, whereas about 4 percent of the blacks and Hispanics were in this group. Similarly, about twice as many Hispanics and blacks reported that their parents had not graduated from high school as the nation as a whole. The ratio was reversed for students who reported that their parents had completed some education beyond high school.

Change in Performance

In every assessment, between half and two-thirds of the exercises are not released so that they can be readministered in subsequent assessments to measure change in performance over time. These exercises are always administered using identical administration and scoring procedures. Consequently, the second mathematics assessment provides a carefully documented measure of change in performance from the first mathematics assessment administered five years earlier.

The change results are summarized in Table A.2. At each age level, the difference between the black average and the national average decreased from 1973 to 1978, although the decrease was small at the older ages. In 1973, black 9-year-olds averaged 15 percentage points below the national average for 9-year-olds. In 1978, this difference had been reduced to 11 percentage points. Thus at age 9 the difference between the black and national averages was reduced by 4 percentage points from the first to the second mathematics assessment. At age 13 the difference was narrowed by 2 percentage points and at age 17, by 1 percentage point. Whereas the national average declined by 1 percentage point at age 9, the black average increased by 3 points. At age 13, the national average declined 2 percentage points, and the black average remained constant. At age 17, performance of both groups declined. Hispanic performance showed a slightly different pattern. Hispanics improved their relative performance by 2 percentage points at ages 9 and 17, but dropped 1 point at age 13.

Table A.2

Change Results

| | Mean Percent Correct for All Change Exercises Administered | | | | | |
| | Age 9 | | Age 13 | | Age 17 | |
	1973	1978	1973	1978	1973	1978
Nation	38	37	53	51	52	48
Black	23	26	32	32	34	31
Hispanic	28	29	40	37	38	36

Some caution is necessary in interpreting the significance of these changes. Some fluctuation is bound to occur over a five-year period. There was a reasonably consistent pattern of improvement over all age groups for both blacks and Hispanics, but given the magnitude of the differences to be overcome, the gains were relatively modest. Only the black 9-year-olds improved their relative performance by more than 2 percentage points. However, these gains, while not large, should not be minimized. The 4-percentage-point gain of black 9-year-olds is over a fourth of the difference that existed between the black and the national average for 9-year-olds in 1972.

Analysis of Achievement by Content Area and Cognitive Level

The assessment focused on five major content areas: (1) numbers and numeration, (2) variables and relationships, (3) geometry (size, shape, and position), (4) measurement, and (5) other topics, which included probability and statistics and graphs and tables. Each content area was assessed at four cognitive levels: (1) knowledge, (2) skill, (3) understanding, and (4) application. Knowledge level exercises involved the recall of facts and definitions. This included such tasks as ordering numbers; recalling basic addition, subtraction, multiplication, and division facts; identifying geometric figures; and identifying basic measurement units. Skill exercises involved various mathematical manipulations including computation with whole numbers, fractions, decimals, and percents. Also included were making measurements, converting measurement units, reading graphs and tables, and manipulating algebraic expressions. Understanding exercises tested students' knowledge of such basic underlying principles as the concept of a unit in measurement. These exercises were constructed so that students could not simply apply a routine algorithm. Application exercises required students to apply knowledge or skills to solve a problem. Both routine textbook problems and nonroutine problems were included in this category.

The difference between the national average and the averages of blacks and Hispanics was remarkably consistent over all four cognitive levels. For example, at age 9, black students averaged 11 percentage points below the national average in the knowledge, skills, and applications categories and 10 points below on the exercises measuring understanding. Similar patterns were found over all three age groups for both blacks and Hispanics. The difference scores for the four cognitive levels never deviated by more than 3 percentage points, and in every case but one they were only a point or two apart. Thus, no cognitive level was disproportionately more difficult for blacks and Hispanics than it was for the nation as a whole.

Performance in specific content areas was not as consistent, although the performance of blacks and Hispanics generally followed the performance pattern of the nation. There was some evidence that measurement topics were disproportionately difficult for black students, but in general it appears that blacks and Hispanics were successful in topics in which all students were successful and had the same areas of difficulty as students in general. These data suggest that they were reasonably good at computation, especially computation involving whole numbers. But they have not learned many basic mathematics concepts in such areas as geometry, measurement, reading graphs and tables, and probability and statistics. Hispanic and black students also share with all students the serious need to develop a deeper understanding of many basic mathematics concepts and to improve problem-solving skills.

Course Background

The National Council of Teachers of Mathematics (1980) has proposed that at least three years of mathematics be required of all students in grades 9 through 12. The assessment's course-background data summarized in Table A.3 indicate that current enrollment is far short of that goal. The majority of 17-year-olds have taken only two years of high school mathematics. Black enrollment in high school mathematics courses falls even further short. Blacks appear to take about one year less of high school mathematics than their peers. It is clear that participation in advanced mathematics courses is an area in which serious inequities exist for black students.

These course-participation data may help to explain why the difference between black achievement and the national average increases with age. As the data in Table A.4 clearly demonstrate, performance was directly related to the amount of mathematics studied. For each additional course taken, there was a significant increase in the level of performance for both black and national averages. These data do not conclusively demonstrate a cause-and-effect relationship between the number of additional mathematics courses taken and achievement. However, when course background was held constant, the difference between the black and national averages was considerably less than the overall difference for all 17-year-olds. The only exception to this rule occurred for algebra 2. Why this course should be an exception is not clear. It is possible that there was more error in classifying students in algebra 2. All course-background data were based on self-report. Many schools teach an introductory algebra course over a two-year period. Some students who had taken this course may have responded that they had taken a second year of algebra.

Unfortunately, comparable course-background data for Hispanics were not summarized by National Assessment. Although it was possible to generate reasonably reliable averages for Hispanics when a large number of exercises were combined to calculate achievement averages, the size of the sample did not generate reliable data for individual questions. For the same reason, affective data are not available for Hispanics.

Table A.3

Mathematics Course Background of 17-Year-Olds[*]

| Course | Percent of 17-Year-Olds Who Have Taken at Least One-Half Year | |
	Nation	Black
Algebra 1	72	55
Geometry	51	31
Algebra 2	37	24
Trigonometry	13	7

[*]Participation in prealgebra, general mathematics, and precalculus classes was also surveyed. However, there appears to be the possibility of significant error in these measures, and so they are not included in this table.

Table A.4

Achievement by Course Background for 17-Year-Olds

| Level | Mean Percent Correct | | | | |
	All 17-Year-Olds	< Alg. 1	Alg. 1	Geom.	Alg. 2
Skills					
Nation	59	42	53	61	70
Black	41	33	42	47	52
Understanding					
Nation	58	42	51	62	67
Black	41	34	40	47	48
Applications					
Nation	44	29	36	46	52
Black	26	21	25	30	30

Affective Data

In contrast to the achievement results and course-background data, the affective results provide a positive picture of black students' feelings about mathematics and themselves as learners of mathematics. In general, their affective responses compared favorably with those of their peers.

The assessment's affective measures are subject to a number of threats to their validity. Although this problem is serious, the data still provide a valuable perspective on black students' attitudes toward mathematics and themselves as learners of mathematics. Unlike most attempts to measure affective variables, the assessment exercises are not aggregated into scales. Instead, they are reported on an exercise-by-exercise basis, and it is not reasonable to calculate averages even for comparison purposes. Consequently, results for selected exercises are reported below.

The results summarized in Table A.5 indicate that most black students liked mathematics and thought it was important. In fact, at every age, black students rated mathematics as the most important school subject. Black students not only responded that they liked mathematics and thought it was important, they also indicated that they wanted to do well in mathematics and were willing to work hard to do so. Over 90 percent of the black 13- and 17-year-olds agreed with each of the following statements:

> I really want to do well in mathematics.
> My parents really want me to do well in mathematics.
> I am willing to work hard to do well in mathematics.
> A good grade in mathematics is important to me.

Table A.5

Rating of Mathematics as a School Subject

	Like	Dislike	Percent Responding* Easy	Hard	Important	Unimportant
Age 9						
Nation	65	11	42	13	**	**
Black	69	11	50	16	**	**
Age 13						
Nation	69	22	56	29	90	4
Black	69	20	60	22	82	5
Age 17						
Nation	59	32	48	36	88	6
Black	64	23	44	42	87	5

*Percentages do not total 100 because of the undecided and "did not take course" categories.
**Not administered at age 9

Black students also responded that they wanted to take more mathematics and even enter a career using mathematics (Table A.6). The results summarized in Table A.6 provide a stark contrast to the course-background data summarized in Table A.3. Although black students indicated a greater desire than their peers to take more mathematics, they actually took fewer advanced mathematics courses.

Conclusions

The assessment results clearly document the serious inequities that exist in the mathematics education of black and Hispanic students. The results also suggest that the achievement of blacks and Hispanics will respond as the inequities are addressed. It appears that some improvement occurred in the status of black and Hispanic education in mathematics between the first and second mathematics assessment, at least at the primary level. During this period, a wide range of federal, state, and local programs was instituted to improve the education of minority and economically disadvantaged students. Some of these programs may be paying off. However, the assessment data provide no hint of what specific factors contributed to the achievement gains.

Although the assessment results do not help spell out potential programs of action, they do help to define the scope and dimensions of the prob-

Table A.6

Further Study of Use of Mathematics

| Statement | Percent Responding[*] | |
	Agree	Disagree
I would like to take more mathematics.		
Age 13		
Nation	49	25
Black	67	20
Age 17		
Nation	39	32
Black	47	23
I would like to work at a job using mathematics.		
Age 9		
Nation	44	22
Black	50	24

[*]Percentages do not total 100 because of the undecided category.

lem. They suggest that motivation may not be a major hurdle. Black students appear to recognize the importance of mathematics and are motivated to continue studying mathematics and be successful in it. The course-background data gathered in the assessment indicate that there is a serious inequity in enrollment in mathematics courses. There needs to be a conscious effort to increase minority participation in advanced mathematics classes.

Finally, the assessment results indicate that black and Hispanic students' strengths and weaknesses parallel those of students in general. These results appear to have definite implications for instruction. The goals for improving minority achievement should not be based on current standards of achievement in the nation. They should be based on what should be learned, not on what is currently being learned. Consequently, programs directed at improving the performance of minority students need to provide more than routine drill on rote computation. They need to focus on a wide range of basic skills, to develop understanding of fundamental mathematics concepts, and provide ample opportunity for a variety of problem-solving activities.

The recommendations urged in An Agenda for Action: Recommendations for School Mathematics of the 1980s (National Council of Teachers of Mathematics, 1980) address a number of serious issues in the mathematics curriculum. The National Assessment results suggest that providing real equity for minorities in the learning of mathematics is as critical an issue as any identified in the NCTM recommendations. These recommendations are valid for all students and should serve as the basis for improving minority achievement in mathematics. A major goal for the 1980s should be to eliminate the inequities that exist in the mathematics education of minority students.

Reference

National Council of Teachers of Mathematics. An Agenda for Action: Recommendations for School Mathematics of the 1980s. Reston, Va.: The Council, 1980.

Appendix B
Sex-related Differences
in Mathematics

Elizabeth Fennema

Thomas P. Carpenter

The second mathematics assessment of the National Assessment of Educational Progress provides new insight into the problems of sex-related differences in mathematics. Information about courses taken and achievement in specific content areas and at different cognitive levels is available from a representative national sample of over 70 000 9-, 13-, and 17-year-olds. The purpose of this article is to report the sex-related differences that were found in this assessment and to explore the significance of these differences.

Participation in Mathematics Courses

Table B.1 shows the percentages of 17-year-old females and males who reported that they had been enrolled for at least half a year in specific mathematics courses. While there are a number of questions that could be raised about the reliability of these data, there was probably little systematic variation by sex in response. These data indicate that there is very little difference between females and males in the mathematics courses taken in the early portion of the high school mathematics sequence.

Table B.1
Mathematics Course Background

| Course | Percent of 17-Year-Olds Who Have Taken at Least 1/2 Year | |
	Females	Males
General or Business Mathematics	47	44
Prealgebra	45	46
Algebra 1	74	71
Geometry	51	52
Algebra 2	36	38
Trigonometry	11	15
Precalculus/Calculus	3	5

It is more difficult to interpret the data for the advanced courses. The majority of students were assessed in the spring of their junior year and would have had limited opportunity to take a fourth year of mathematics. Thus, no clear conclusions can be drawn regarding participation in advanced mathematics courses. The data that do exist, however, suggest that sex-related differences in course participation emerge in the most advanced mathematics courses. Although the absolute differences in the percentages of females and males who took trigonometry and precalculus/calculus were not large, there were relatively few students of either sex who reported they took these courses. Only about two-thirds as many females as males reported they had taken either of these courses. If this ratio is representative of the relative enrollment of females in advanced mathematics courses, these findings would help explain

the relatively low number of females in careers or university majors that require a strong background in advanced mathematics.

Sex Differences in Achievement

Assessment exercises measuring achievement were categorized by mathematical content and cognitive level. Five content areas were assessed: number
and numeration, variables and relationships, geometry, measurement, and other
topics. Each content area was assessed at four cognitive levels: knowledge,
skill, understanding, and application. The scores for the sets of items representing the four cognitive levels assessed are summarized in Table B.2. On
each cluster of items the percent of correct responses for males was subtracted from the percent of correct responses for females. A positive score indicates that females scored higher than males; a negative score indicates the
reverse.

Table B.2

Differences between Female and Male Achievement

Cognitive Level	Difference between Female and Male Average Percent Correct*		
	Age 9	Age 13	Age 17
Knowledge	1.41	−.17	−2.16
Skills	.40	1.11	−2.54
Understanding	−1.07	−.29	−3.61
Applications	−.37	−1.60	−5.04

*Calculated by subtracting the average male score for all exercises in the
given category from the average female score. A positive score represents
higher female performance; a negative score, higher male performance.

No clear pattern of differences in achievement is apparent at ages 9 or
13. Averages of females tended to be slightly higher on the knowledge and
skills exercises and those of males somewhat higher for understanding and
applications. At age 17, males' average performance exceeded that of females
at every cognitive level.

Another way to look at achievement data is to compare the achievement of
17-year-old females and males who reported that they had been enrolled in the
same mathematics courses. Table B.3 shows these data. The results in Table

Table B.3

Differences between Female and Male Achievement at Age 17

Cognitive Level	Difference between Female and Male Average Percent Correct*				
	<Alg. 1	Alg. 1	Geom.	Alg. 2	>Alg. 2
Knowledge	−.6	−2.2	−2.6	−2.2	−3.1
Skills	−.64	−2.32	−2.39	−3.29	−3.79
Understanding	−1.75	−2.59	−3.83	−4.25	−4.71
Applications	−2.44	−4.40	−4.55	−6.33	−7.41

*Calculated as in Table B.2

B.3 indicate that achievement differences still exist when course background was taken into account. For each course-background category, male achievement exceeded that of females. It should also be noted that the magnitude of the difference increased consistently in relation to the amount of mathematics taken. In other words, the difference in performance between 17-year-old females and males was smallest for students who had not taken first-year algebra, greater for students who had taken first-year algebra, and even greater for students who had taken geometry; the trend continued through courses beyond second-year algebra.

Another consistent trend was observable in the 17-year-old achievement results. The achievement differences between females and males increased with the cognitive level. There were smaller differences at the lower cognitive levels and larger differences at the higher levels.

Differences in Specific Areas of Achievement

Within specific content areas, different patterns of performance emerged. Females scored higher on lower-level number and numeration skills at ages 9 and 13. Males scored higher on multistep word problems in this content area at all ages (Table B.4). A different pattern of results was found on geometry and measurement exercises (Table B.5). At ages 9 and 13, there was a consistent pattern of lower averages for females on geometry and measurement exercises over all cognitive levels. For measurement, these differences were often substantial.

Table B.4

Performance on Number and Numeration Exercises

Cognitive Level	Difference between Female and Male Average Percent Correct*		
	Age 9	Age 13	Age 17
Knowledge	2.28	1.16	-.58
Computation Skills	2.36	3.51	-.93
Understanding	-1.47	1.32	-1.60
One-Step Word Problems**	.28	-.51	-5.16
Multistep Word Problems**	-2.25	-3.40	-4.97

*Calculated as in Table B.2
**The applications level was divided into these two categories.

A possible explanation for females' relatively poor performance on geometry and measurement exercises is that a substantial number of these exercises may involve spatial visualization skills. From about adolescence, females perform at lower levels than males on items that measure this skill (Maccoby and Jacklin, 1974). Several geometry exercises in the assessment appeared to require a direct application of spatial visualization skills, and spatial visualization skills may have played a part in the solution of many other geometry and measurement exercises. An example of an assessment exercise that seems to require direct application of spatial visualization is presented in Figure B.1.

At age 13, females scored more than 6 percentage points lower than males on this exercise, and at age 17, the difference was more than 16 percentage points. Although the results on other exercises apparently involving spatial visualization seem to support the hypothesis that spatial visualization skills play a key role in sex-related differences in mathematics, the results were

Table B.5

Performance on Geometry and Measurement Exercises

Category	Difference between Female and Male Average Percent Correct*		
	Age 9	Age 13	Age 17
Geometry			
Knowledge	-.62	-.31	-2.78
Skills	-1.76	-1.38	-6.07
Understanding	-1.26	-1.45	-4.62
Applications	**	-1.91	-6.35
Measurement			
Knowledge	-1.88	-6.51	-7.17
Skills	-2.06	-4.76	-9.58
Applications	-4.32	-5.55	-7.64

*Calculated as in Table B.2
**Not available for age 9

Finish drawing the figure on the right so
it is congruent to the figure on the left.

Figure B.1

not consistent. No differences were found on a number of exercises that also
seem to require a direct application of spatial visualization skills. Furthermore, some of the largest differences in geometry and measurement exercises
were found on exercises that seemed to depend very little on spatial visualization. For example, 13-year-old females scored 11 percentage points lower than
males on the following exercise:

> Mr. Jones put a wire fence all the way around his rectangular
> garden. The garden is ten feet long and six feet wide. How
> many feet of fencing did he use?

Although constructing a picture or mental image of a rectangle might help solve
this problem, almost all 13-year-olds could draw a rectangle, and so spatial
skills would not seem to be a limiting factor in this case. Overall, the
assessment results appeared to provide no clear resolution of the question of
what role spatial visualization skills play in sex differences in mathematics
achievement.

Change in Performance

A number of exercises were given in both the 1973 and 1978 assessments. Results of the performance on these items furnish a measure of change or stability over time. In 1973, no information was gathered about course background, and so the samples from which these data were drawn were not controlled for course background. The change results are summarized in Table B.6.

Table B.6

Change Results

Cognitive Level	Difference between Female and Male Average Percent Correct for Change Exercises*		
	Age 9	Age 13	Age 17
Knowledge and Skills			
1973	.21	.90	-3.19
1978	.04	-1.8	-2.69
Understanding			
1973	.48	-2.38	-4.78
1978	.12	-1.73	-4.43
Applications			
1973	-1.82	-2.68	-6.53
1978	-1.82	-1.78	-5.02

*Calculated by subtracting the average male score for all change exercises in the given category from the average female score. A positive score represents higher female performance; a negative score, higher male performance. These figures are slightly different from those reported in Table B.2 because only change exercises are involved in this calculation.

At ages 9 and 13, patterns of sex differences were less clear to begin with and there was not as clear a pattern of change. The relative performance of females improved in some categories and declined in others. The overall difference between female and male performance at age 17 decreased from the first assessment to the second. On the second assessment, the overall difference in performance between female and male 17-year-olds averaged 0.68 percentage points less than it did on the first assessment; for the application exercises, the difference was reduced by 1.51 percentage points.

Conclusions

The assessment results indicate that on a nationwide basis there is little difference between males and females in overall mathematics achievement at ages 9 and 13. At age 17, however, females are not achieving at the same level in mathematics that males are. Even when females and males reported that they had been enrolled in the same mathematics courses, males' performance was higher than that of females, and the differences were greatest on the more complex tasks.

We believe that these results are indicative of a continuing inequity in mathematics education. It has been hoped that such differences in achievement between females and males would disappear when there was equal enrollment in elective mathematics courses. Such is not the case. These results suggest that attaining equal achievement in mathematics is a complex problem that requires more than ensuring that females enroll in mathematics courses. While encouraging females to enroll in advanced courses is an important first step, it is not sufficient. To say that increasing females' enrollment in mathematics is not sufficient does not imply that enrollment in mathematics courses is unrelated to achievement. In fact, the assessment change data that show a relative improvement of females' achievement at age 17 may reflect greater participation by females in mathematics courses over the last few years.

In its Agenda for Action the National Council of Teachers of Mathematics (1980) provides a broad set of recommendations for redirecting the mathematics curriculum of all students. The assessment results clearly document that we are currently far short of providing the type of education envisioned in this proposal (Carpenter, Corbitt, Kepner, Lindquist, and Reys, forthcoming). These recommendations for the mathematics curriculum of the 1980s propose that students should take at least three years of high school mathematics. The assessment results indicate that fewer than 40 percent of the 17-year-olds of either sex are reaching that goal. The results also document that performance of all students is inadequate in many areas, especially those in which problem solving or applications of mathematical skills are involved (Carpenter et al., forthcoming). We believe that in working to overcome these broadly based deficiencies, specific attention must be given to improving achievement for females. Focusing exclusively on overcoming broadly based deficiencies probably will not eliminate the inequitable outcomes that have traditionally existed. Individual school systems need to implement programs that are designed to eliminate achievement differences between males and females.

Unfortunately, the mathematics assessment results, while documenting the problem of sex-related differences in mathematics more precisely than has been done before, offer little help in identifying the causes of these differences (when such identification could lead to effective intervention procedures). Reasons for the differences and possible intervention procedures have been addressed elsewhere (Fennema, 1980; Fennema, Wolleat, Pedro, and Becker, 1981; Jacobs, 1978). Concerned school systems and teachers should seek out information about such programs and procedures, or they should develop their own programs. Such programs do make a difference, and equitable education for males and females can be achieved.

References

Carpenter, T. P., M. K. Corbitt, H. S. Kepner, Jr., M. M. Lindquist, and R. Reys. "National Assessment: Implications for the Curriculum of the 1980s." In Implications of Research in Mathematics Education for the Curriculum of the 80s, edited by E. Fennema. Washington, D.C.: Association for Supervision and Curriculum Development, forthcoming.

Fennema, E. "Teachers and Sex Bias in Mathematics." Mathematics Teacher 73 (March 1980): 169-73.

Fennema, E., P. L. Wolleat, J. D. Pedro, and A. D. Becker. "Increasing Women's Participation in Mathematics: An Intervention Study." Journal for Research in Mathematics Education 12 (January 1981): 3-14.

Jacobs, J. E., ed. Perspectives on Women and Mathematics. Columbus, Ohio: ERIC Clearinghouse for Science, Mathematics, and Environmental Education, 1978.

Maccoby, E. E., and C. N. Jacklin. The Psychology of Sex Differences. Stanford, Calif.: Stanford University Press, 1974.

National Council of Teachers of Mathematics. An Agenda for Action: Recommendations for School Mathematics of the 1980s. Reston, Va.: The Council, 1980.

References

Second Mathematics Assessment
1977-78

NAEP Publications
(Available from NAEP, Suite 700, 1860 Lincoln Street, Denver, CO   80295)

National Assessment of Educational Progress. <u>Changes in Mathematical Achieve-</u>ment: 1973-78. 09-MA-01. August 1979.

_____. <u>Mathematical Applications</u>. 09-MA-03. August 1979.

_____. <u>Mathematical Knowledge and Skills</u>. 09-MA-02. August 1979.

_____. <u>Mathematical Objectives</u>: Second Assessment. Denver: NAEP, 1978.

_____. <u>Mathematical Understanding</u>. 09-MA-04. December 1979.

_____. <u>The Second Assessment of Mathematics, 1977-78: Released Exercise Set</u>. Denver: NAEP, May 1979.

NCTM Publications

Anick, Constance M., Thomas P. Carpenter, and Carol Smith. "Minorities and Mathematics: Results from National Assessment." <u>Mathematics Teacher</u>, in press.

Bestgen, Barbara J. "Making and Interpreting Graphs and Tables: Results and Implications from National Assessment." <u>Arithmetic Teacher</u> 28 (December 1980): 26-29.

Carpenter, Thomas P., Mary Kay Corbitt, Henry S. Kepner, Jr., Mary Montgomery Lindquist, and Robert E. Reys. "Calculators in Testing Situations: Results and Implications from National Assessment." <u>Arithmetic Teacher</u> 28 (January 1981): 34-37.

_____. "The Current Status of Computer Literacy: NAEP Results for Secondary Students." <u>Mathematics Teacher</u> 73 (December 1980): 669-73.

_____. "Decimals: Results and Implications from National Assessment." <u>Arithmetic Teacher</u> 28 (April 1981): 34-37.

_____. "NAEP Note: Problem Solving." <u>Mathematics Teacher</u> 73 (September 1980): 427-33.

_____. "Results and Implications of the Second NAEP Mathematics Assessment: Elementary School." <u>Arithmetic Teacher</u> 27 (April 1980): 10-12, 44-47.

_____. "Results of the Second NAEP Mathematics Assessment." <u>Mathematics Teacher</u> 73 (May 1980): 329-38.

_____. "Solving Verbal Problems: Results and Implications from National Assessment." <u>Arithmetic Teacher</u> 28 (September 1980): 8-12.

_____. "Students' Affective Responses to Mathematics: Results and Implications from National Assessment." <u>Arithmetic Teacher</u> 28 (October 1980): 34-37, 52-53.

_____. "Students' Affective Responses to Mathematics: Secondary School Results from National Assessment." <u>Mathematics Teacher</u> 73 (October 1980): 531-39.

_____. "What Are the Chances of Your Students Knowing Probability?" <u>Mathematics Teacher</u> 74 (May 1981): 342-44.

Fennema, Elizabeth, and Thomas P. Carpenter. "Sex Differences in Mathematics: Results from National Assessment." <u>Mathematics Teacher</u>, in press.

Hiebert, James. "Units of Measure: Results and Implications from National Assessment." <u>Arithmetic Teacher</u> 28 (February 1981): 38-43.

Hirstein, James. "The Second National Assessment in Mathematics: Area and Volume." <u>Mathematics Teacher</u>, in press.

Kerr, Donald. "A Geometry Lesson from National Assessment." <u>Mathematics Teacher</u> 74 (January 1981): 27-32.

McKillip, William D. "Computational Skill in Division: Results and Implications from National Assessment." <u>Arithmetic Teacher</u> 28 (March 1981): 34-37.

Post, Thomas R. "Fractions: Results and Implications from National Assessment." <u>Arithmetic Teacher</u> 28 (May 1981): 26-31.

Rathmell, Edward C. "Concepts of the Fundamental Operations: Results and Implications from National Assessment." <u>Arithmetic Teacher</u> 28 (November 1980): 34-37.

Other Publications

Carpenter, T. P., M. K. Corbitt, H. S. Kepner, Jr., M. M. Lindquist, and R. E. Reys. "National Assessment: Implications for the Curriculum of the 1980s." In <u>Research in Mathematics Education: Implications for the 80s</u>, edited by Elizabeth Fennema. Washington, D.C.: Association for Supervision and Curriculum Development, forthcoming.

_____. "A Perspective of Students' Mastery of Basic Skills." In <u>Selected Issues in Mathematics Education</u>, edited by Mary Montgomery Lindquist. Chicago: National Society for the Study of Education; Reston, Va.: National Council of Teachers of Mathematics, 1981.

_____. "Problem Solving in Mathematics: National Assessment Results." <u>Educational Leadership</u> 37 (April 1980): 562-63.

First Mathematics Assessment
1972-73

NAEP Publications

National Assessment of Educational Progress. <u>Consumer Math: Selected Results from the First National Assessment of Mathematics</u>. 04-MA-02. June 1975.

_____. _The First National Assessment of Mathematics: An Overview_. 04-MA-00. October 1975.

_____. _Math Fundamentals: Selected Results from the First National Assessment of Mathematics_. 04-MA-01. January 1975.

_____. _Mathematics Objectives_. Ann Arbor, Mich.: NAEP, 1970.

_____. _Mathematics Technical Report: Exercise Volume_. 04-MA-20. February 1977.

_____. _Mathematics Technical Report: Summary Volume_. 04-MA-21. September 1976.

NCTM Publications

The following were prepared by the NCTM Project for Interpretive Reports team: Thomas P. Carpenter, Terrence G. Coburn, Robert E. Reys, and James W. Wilson.

"Notes from National Assessment: Addition and Multiplication with Fractions." _Arithmetic Teacher_ 23 (February 1976): 137-42.

"Notes from National Assessment: Basic Concepts of Area and Volume." _Arithmetic Teacher_ 22 (October 1975): 501-7.

"Notes from National Assessment: Estimation." _Arithmetic Teacher_ 23 (April 1976): 296-302.

"Notes from National Assessment: Perimeter and Area." _Arithmetic Teacher_ 22 (November 1975): 586-90.

"Notes from National Assessment: Processes Used on Computational Exercises." _Arithmetic Teacher_ 23 (March 1976): 217-22.

"Notes from National Assessment: Recognizing and Naming Solids." _Arithmetic Teacher_ 23 (January 1976): 62-66.

"Notes from National Assessment: Word Problems." _Arithmetic Teacher_ 23 (May 1976): 389-93.

"Research Implications and Questions from the Year 04 NAEP Mathematics Assessment." _Journal for Research in Mathematics Education_ 7 (November 1976): 327-36.

"Results and Implications of the NAEP Mathematics Assessment: Elementary School." _Arithmetic Teacher_ 22 (October 1975): 438-50.

"Results and Implications of the NAEP Mathematics Assessment: Secondary School." _Mathematics Teacher_ 68 (October 1975): 453-70.

Results from the First Mathematics Assessment of the National Assessment of Educational Progress. Reston, Va.: National Council of Teachers of Mathematics, 1978.

"Subtraction: What Do Students Know?" _Arithmetic Teacher_ 22 (December 1975): 653-57.

Other References

Martin, Wayne L., and James W. Wilson. "The Status of National Assessment in Mathematics." Arithmetic Teacher 21 (January 1974): 49-53.

Reys, Robert E. "Consumer Math: Just How Knowledgeable Are U.S. Young Adults?" Phi Delta Kappan (November 1976): 258-60.